– Locks and Keys throughout the Ages –

Locks and Keys throughout the Ages

Vincent J. M. Eras

Published by Artisan Ideas, 2019.
Text by Vincent J. M. Eras.

ISBN 978-0-9979798-6-2
Library of Congress Control Number: 2019945545.
Printed in China.
Artisan Ideas is an imprint of **Artisan North America, Inc**
Tel: 800-843-9567
Info@ArtisanIdeas.com
Copyright © 1957 by Vincent J. M. Eras, Lips' Safe and Lock Manufacturing Company, and ASSA ABLOY, under International Copyright Union and U.S.A. All rights reserved.

For a vast selection of reference books for blacksmiths and bladesmiths visit our website: www.ArtisanIdeas.com

All rights reserved. No part of this publication may be reproduced, distributed, or transmitted in any form or by any means, including photocopying, recording, or other electronic or mechanical methods, without prior written permission of the publisher, except in the case of brief quotations embodied in critical reviews and certain other noncommercial uses permitted by copyright law.
For permission requests, write to the pubblisher, addressed "Attention: Permissions Department", at the address below.
The scanning, uploading, and distribution of this book or any part thereof via the internet or via any oiother means without the written permission of the publisher is illegal and punishable by law.

**Artisan North America
753 Valley Road
Watchung, NJ 07069-6120**

Ordering information:
Quantity sales. Special discounts are available on quantity purchases by corporations, associations, and others. For details, contact the publisher at the address above.
For single copies, please try your bookstore first. If unavailable this book can be ordered through: www.ArtisanIdeas.com.
To see our complete selection of books and DVDs on this and related subjects visit our website listed above.

Authors!

If you have an idea for a book please contact us at: Info@ArtisanIdeas.com
We'll be glad to hear from you.

Front cover photo: from the Lips' collection, courtesy of ASSA ABLOY, Holland.

www.ArtisanIdeas.com

TABLE OF CONTENTS

Introduction to the 2019 printing .... (7)
Publisher's Notes Regarding the 2019 Printing .... (8)
Author's Preface .... (10)
Original Introduction .... (14)
Terminology .... (16)
Etymology of the Words, Locks and Keys .... (21)

CHAPTER I   *What is a Lock?* .... (24)
The Gordian knot - Egyptian, Roman and Greek "Safe Deposit" vaults - The Egyptian lock - The Great Temple of Karnak - A Moroccan lock - Tanimbar Island lock - A Maroon lock from Surinam - A lock from the Faroe Islands

CHAPTER II   *Metallurgy* .... (35)
"Le parfait Serrurier" - Ancient Egyptian keys - Ancient Roman keys (George Price) - The oldest Roman and Greek keys (Lips' Collection) - Keys in Dechelette's Collection - Greek temple keys - Keys symbolized in Coats of Arms and Seals - Ancient padlocks - Oriental padlocks - Locks and keys in the Middle Ages - Sliding keys - Roman, Frankish, Merovingian and Karolingian keys - Ancient keys in the Lips' Collection - Some groups of ornamental keys of the Lips' collection - Some groups of keys from the collection "Le Secq des Tournelles" - Specimens of French art: Musee de Cluny, Paris - Ornamental French keys: Museo Nazionale. Florence.

CHAPTER III   *The First Metal Locks* .... (65)
Iron keys with copper ornamental bows: Städtisches Museum - A locksmith's bill (1709) - Collection Frans Engels, Antwerp, Belgium - Collection Dr. E. Vita Israëls, Amsterdam, Holland - Collection Museum Princessehof, Leeuwarden, Holland - Australian keys presented by Messrs. Wm. Bedford Limited, Melbourne - The key of St. Gervase of Maastricht - English ornamental keys - Ancient ornamental locks preserved in the Musee de Cluny, Paris, and the Museo Artistico Industriale, Rome - An old lockmaker's tool presented by Mr. J. D. Cartsens, Deventer, Holland - Ancient locks collected in Museums in Germany, France, England, Spain and other countries - Ancient locks in the Lips' Collection - Ancient locks collected in the Bayerisches National Museum, Munich - Chest lock in the Städtisches Museum, Nürnberg - The lock of the main entrance door of Mons Town Hall - Collection in the Rijks Museum, Amsterdam - Old Italian and Swiss locks in the Lips' Collection - Ancient padlocks in the Lips' Collection - Old Chinese safe.

CHAPTER IV   *Locks and Keys and their Manufacture* .... (102)
Warded locks of European origin - Old locks and their bushed keyholes - Warded locks.

CHAPTER V	*Modern Locks and their Applications* .... **(109)**

Lever locks - Barron's twin tumbler lock - Bramah's lock - The Bramah-Vago lock - The Bramah-Chubb system - Cotterill's Climax-Detector lock - Chubb's lever lock - Levers - The Chubb Detector - Lips' levers - Cylinder locks - Lips' drillproof cylinders - Lock picking - Hobbs' lock picking tools - The writer and his modern lock picking tools.

CHAPTER VI	*Permutations in Locks and Keys* .... **(126)**

Combinations - Modern locks and their uses - Cupboard and room door locks - Lever handle and knob operated locks - Front door lever and cylinder locks - Master keyed lever locks - Master keyed cylinder locks - Fixing of locks and striking plates.

CHAPTER VII	*Locks for Safes and Strong Room Doors* .... **(138)**

The three and four dial letter lock - Bauche's lock - The Bramah-Chubb lock - The French letter combination click lock - The French letter repeater click lock - Kromer's protector lock - Lips' key operated safe door lock - Dr. Andrew's key-operated safe door lock - Day and Newell's key operated safe door lock - Yale's double treasury lock - Yale's double Quadruplex lock - Herring's Grasshopper lock - Isham's key and combination lock.

CHAPTER VIII	*Safe Deposit Box Locks* .... **(149)**

Lips' changeable lever lock - Lips' interchangeable key operated lock - Lips' springless lever lock - Lips' key and combination lock combined - Yale's cylinder safe deposit box lock - Yale's changeable safe deposit box lock - Rosengren's safe deposit box lock - Mr. E. J. Goodnough on combination locks.

CHAPTER IX	*Some General Remarks on Keyless Combination Locks* .... **(159)**

Letter combination locks - Kromer's letter combination lock - Cipher combination locks - Sargent's cipher combination lock - Sargent's Micrometer - Pillard's cipher combination lock - Yale's double dial pull combination lock - Dexter's double dial combination lock - Jones' patent combination lock - Dexter's single dial combination lock - Lillie's patent combination lock - Lips' cipher combination lock - Chronometer time locks - Sargent's first patented time lock - Pillard's time lock - Holmes' electric time lock - Dalton's dual time lock - Automatic locking devices and time lock combined - Lips' 40-ton circular vault door

Introduction to the 2019 Printing

This wonderful book again proves that the past is a motivator for the present. Vincent J.M. Eras clearly describes the development of locks and keys throughout the ages. Be guided by this history and realize that using the latest technologies and tried-and-tested techniques, the world of locks and keys is constantly developing innovative solutions that improve customers' lives through greater security, safety and convenience.

ASSA ABLOY, the global leader in access solutions, is pleased that **Artisan Ideas** has re-published this book and wishes you a lot of reading pleasure.

Publisher's Notes Regarding the 2019 Printing

"*Locks and Keys throughout the Ages*" was originally published by the **Lips' Safe and Lock Manufacturing Company**, Dordrecht, Holland, in 1957.

It was republished in the United Kingdom in 1974 by Bailey Brothers and Swinfen.

For this new printing, done with the kind permission of **ASSA ABLOY**, the original typeface and the fonts of the text have been changed for easier reading.

The 'thousands' separator in writing numbers has been changed from the 'period' or 'full-stop' used in continental Europe to the style used in the USA and the UK which employs a comma. Thus, for example, the number "one-million" is written 1,000,000 in this printing while in the previous two printings it was written as 1.000.000.

The continental European method for writing decimals, which employs a comma, which was used in the original printing has been changed to the method used in the UK and USA, which employs a 'decimal point'. Thus, for example, the decimal 'five-tenths' which was written as '0,5' in the first two printings, is written as '0.5' in this printing.

No spelling, grammatical or syntactical changes have been made to the text, but in several cases punctuation and hyphenation was added for clarity.

While the typeface in this printing has been completely modernized, reproducing the photographs posed a problem. The original photographs are no longer available and many of the photo reproductions in the second, 1974 printing of this book were not up-to-scratch.

We used the latest and best photo scanning technology to digitize and then correct the photos from the original, and superior, 1957 printing. The photos in this 2019 printing are of the same, or, in many cases, of higher quality than the photos in the original edition thanks to digital correction technology.

ARTISAN IDEAS
2019

Dedicated to the Memory of Mr. J. Lips Bzn. 1847-1921.

**Founder of the Lips' Companies of
Dordrecht - Brussels - Milan.**

Author's Preface (1957)

It is now more than 58 years ago that I had the privilege of meeting the Lips — father and son — with whom I have since been associated.

Mr. J. Lips Bzn. the founder of the Lips' factories — died in 1921 and it is in his memory that I write this book to mark my deep respect and friendship towards this brilliant man. As it is impossible within the scope of this book to do full justice to this outstanding and capable man, this brief biography can only partly convey to the reader the lasting value of his achievements and his accomplishments during a business career which can be regarded as highly successful. He was born in Rotterdam. His father had a smithy, made stoves and did all kinds of smiths' jobs. Now and then small safes were made to order.

The young Lips left Elementary School at the age of 14 and became an apprentice in his father's workshop. In 1871, at the age of 23, he took over Mr. J. van den Boon's smithy, a small workshop in Dordrecht, and thus started out independently. It was in fact a very small workshop, his staff consisting of one man and two boy-apprentices only. He followed in his father's footsteps, turning out mainly heating stoves, cooking ranges and doing repairs, but he soon switched to the production of safes and fire proof doors. As an experienced craftsman and excellent mechanic young Lips quickly won an outstanding reputation in his own town and in the neighbouring districts, so much so that his small shop could no longer carry out orders within a reasonable period.

His good name helped him to finance the building of a factory, the adjacent house doing service as showroom to draw the attention of prospective customers to his products and to consolidate his reputation.

In the meantime he married and his wife played a very active part in the ever-extending business of this new enterprise.

In due course Mr. Lips' eldest son joined the Company as a partner and assisted the energetic manager in his comprehensive task. When Lips junior left school his father sent him to England and Germany, there to study the languages and to acquire the necessary office experience, and when he returned at the age of 16 proved to be a valuable asset to his father.

Irreproachable workmanship revealed by great fire outbreaks established the firm's reputation, resulting in enquiries for safes, doors and special bank appliances from all parts of the country and from abroad.

Messrs. Lips then decided to erect a new and larger factory, and it was at this period that the author joined the firm and eventually became a partner.

We moved into the new factory, which was equipped with machines and tools of the latest design and using modern techniques especially suited to the manufacture of safes and vault doors. The production of ranges and repair work had meanwhile been discontinued by the firm and transferred to another organization. Up to that time locks and keys were bought from an independent company, but these products came to be considered so vitally important that it was decided to take up their manufacture ourselves. By this means an efficient technical control was ensured which enabled us to give a sound guarantee with our products and moreover to dispense with outside supplies. The new Lock Factory, which was completed in 1902, was equipped with special machinery for this work. The author was appointed manager of this new factory and has held the position ever since.

Thanks to the harmonious co-operation and perfect understanding between the managers, the activities of this small factory, started with some young workmen, were gradually extended as were those of the safe department, with the result that the factory had again to be enlarged. Finally an entirely new plant of greater capacity was built and equipped with the latest machinery and tools for the manufacture of high precision locks and their parts.

In the space of 55 years this industry has become one of the leading concerns in this line in Europe. Lips' factories have also been established in Brussels and Milan. The joint staff of these factories now exceeds 1650 employees. There are show-rooms and agencies in the principal European towns and sales executives travel all over the world to find new markets for the Lips' products.

Mr. J. Lips died in 1921 at the age of 74, after 22 years of productive and most cordial co-operation with the author.

The Company has continued its activities directed by his three sons and myself, who form the Board of Directors. Mr. B. Lips Jr died 15 May 1950 (72 years).

All those who had the privilege of co-operating with Mr. J. Lips and of working under his management will remember with admiration his inexhaustible energy, enterprise and enthusiasm and the fine example he gave to all of us by his devotion to duty and love of his work. He always had a keen eye for the well-being of those who worked under his judicious and deliberate guidance. To us — his successors — he used to say '*A proper understanding is the surest road to success*'. And it is an undeniable fact that his personal characteristics paved the way to the present specialized industry of worldwide renown. Today the international trade mark **LIPS** is known all over the world, and stands for high-grade products made to ensure safety, which only the best quality material and workmanship can guarantee.

After my 58 years' activity with this firm I consider it a gratifying task to place on record my experiences and through this book save them from oblivion. At the same time an excellent opportunity is presented to show the reader my collection of antique and modern locks — the tangible result of more than 50 years travelling, searching and study in many countries for markets for our products, which collection may be considered unique and as complete as is possible.

With this book I want to achieve a double purpose — to publish my experiences in the field of locks, keys and allied accessories in the hope that those interested may become better acquainted with the subject — and last but not least, to pay homage to the immortal memory of the Founder of the Lips' Concern, who we all appreciate and deeply esteem.

Vincent Eras.

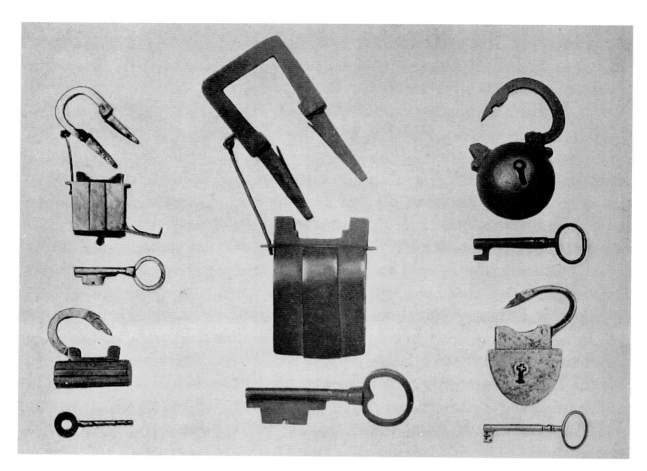

1. A collection of old padlocks presented to the writer by H. M. Queen Wilhelmina of The Netherlands after Her visit to the Lips' Factories.

2. One of the locks to be found in the Royal Palace "Huis ten Bosch" at The Hague, showing the lock of the doors between the Chinese and the Japanese rooms.

3. Key from Tyrol from the writer's Collection.

4. St. George and the Dragon. An English forged steel key dating from the 17th century and one of finest specimens in the Lips' Collection.

Original Introduction

Treatises, encyclopaedias and other literature on ethnology have revealed that the invention and practical use of locks and keys date from many thousands of years before Christ. Locks and keys were used by the Egyptians, the Greeks, the Romans and other nations to safeguard their homes and property. It is surprising, therefore, that in spite of so many centuries of regular use, the knowledge of this device — so familiar to all of us — is still so scanty. We meet it in every walk of life, operate them daily, and yet the number of those whose knowledge extends beyond the simple function of locking and unlocking doors is very small, as compared, for instance, with the general familiarity with motors. Fourteen and fifteen year olds will tell us in detail how a motor works, whether it is a two-stroke or a four-stroke engine, in what way its power is generated, how the petrol is vaporized in the carburettor and properly mixed, how the explosion in the cylinders and transmission of power are effected, and so on.

However, generally speaking, locks should be classified as miniature machines, especially the high grade variety of security locks such as the genuine lever locks, the keyless combination locks — operated by a rotating dial by which certain numbers or letters in a particular order are brought opposite the setting mark — time locks and automatic locks for strong room doors.

After spending a lifetime on the technical development of locks and keys and their manufacture, I could no longer resist the temptation to lift a corner of the veil and bring to light the history of locks and keys and their development throughout the ages, up to the present standard of perfection and refinement achieved through the untiring efforts of modern well equipped factories.

This book does not pretend to be complete in every detail and is bound to give rise to criticism on account of its shortcomings. Not all of the inventions and patents will be discussed, since the principal object is to provide some general information on the achievements in this branch of industry throughout the ages. Assimilation of the contents will be greatly facilitated by the many illustrations, so that by and large I am confident that the book will prove a valuable guide to anyone interested.

I am greatly indebted to my friends — Mr. F. J. Butter of Willenhall, England; Mr. Chas. Courtney of New York, U.S.A.; Mr. Albert H. Hopkins of New York, U.S.A.; Mr. Robert Murray of New York, U.S.A.; Dr. J. A. Molhuysen of The Hague; Mr. J. P. F. H. Kloppenburg of Arnhem; Dr. E. Vita Israël of Amsterdam; Mr. Brian Sixsmith, London; Mr. H. Bainbridge, Shrewsbury; and many others for their valuable information and help, and who, as collectors of old locks and keys, have contributed so much to the instructive character of this record.

Likewise I have to thank Lieut. General Pitt Rivers, F.R.S., Georg Price, Baur, Diels and many other writers, whose books proved such excellent guides on the subject.

When going through this book the reader is invited to discriminate three main sections:

1. The different methods of protecting property – how locks and keys came into being and their development throughout the ages.
2. Modern locks and their general uses.
3. Key operated security locks, keyless combination locks, safe deposit locks, chronometer and time locks and automatic locking devices for strong room doors.

Since there is hardly any literature in this particular field, the writer – by presenting this book – hopes to fill the gap, and invites anyone so inclined to send their criticism.
Finally, I wish particularly to thank Mr. H. H. Fronczek of Amsterdam, who on the basis of his experience as managing editor of Polytechnic-Window of The Netherlands and publisher of this book has devoted his critical attention to it, made the necessary corrections, and through his valuable advice has made its publication possible.

Dordrecht, 1956.   *The author.*

Terminology

A number of terms, expressions and their definitions will be welcomed by the uninitiated, and as space forbids the inclusion of all the names this nomenclature has been adapted to those used in this book:

The Banker's Key:	Used by the custodian to operate one of the locks of a safe-deposit box in the case of a renter's and custodian's control lock.
Bank Lock:	A term for high security locks for safes and strongrooms in banks.
Bit:	The blade or beard projecting from the stem or shank, with the bits and notches, which serves to raise the levers when the key is turned. The number of bits and notches, also called steps, corresponds with the number of levers.
Bolt:	Sliding piece of locks, shooting out of the lock case to fit into a socket or staple on the door frame (see latch and dead bolt).
Case:	The exterior part of a lock in which the mechanisms and action are built.
Change Key Lock:	Operated by any key chosen from a large number of different keys. The selected key is used to throw out the bolt and after that is the only key which will next withdraw it. Changes can be made as often as desired, in excess of one million.
Combination Lock:	One which is operated by a rotating dial by which certain numbers or letters in a particular order, after a given number of turns in the prescribed direction, are brought opposite the setting mark, after which the lock can be opened.
Dial of Combination Lock:	See Combination lock.
Control Key:	Used to re-adjust a combination lock.
Corrugated Key	One of sheet metal in which corrugations are pressed or milled in the bit or shank, as used with cylinder locks.
Cylinder:	The part of a cylinder lock which provides the security. It consists of a short cylindrical plug containing the key hole and mechanism, adjustable by the key.
Cylinder Lock:	One having a cylinder or cylinders with pin tumbler mechanism.
Dead Bolt:	The square or round bolt which is moved both inwards and outwards by the key

Dead Latch:	A latch which can be fixed or is automatically fixed to replace a dead bolt.
Dial Lock:	See Combination lock.
Double Bitted Key:	One with a bit on each side of the shank, with steps, to raise the levers.
Double Throw Lock:	One with a bolt which, after the first throw, can be shot out further by an extra turn of the key, and requires two turns to withdraw it fully.
Drill Pin:	A fixed stump or pin in a lock on which a pipe key fits to rotate.
Escutcheon:	A frontispiece of a lock covering the keyhole.
Espagnolette Bolt:	A door or window fastening having bolts the full height of the door or window which are centrally operated by a central handle or a key.
Flat Key:	One that is made from strip metal and remains flat without groove or corrugation.
Flush Bolt:	A door bolt which can be recessed flush into the edge of a door.
Flush Lock:	A flush fitting lock.
Folding Key:	A key of which the two halves are hinged together and can be folded to facilitate carrying about.
Follower:	The part of the lock which is turned by handle or spindle to withdraw the latch holt.
Gating:	The slot in a lever, through which the bolt stump passes during the travel of dead bolt or runner.
Guard:	A fixed part inside a lock to prevent false keys from turning or to prevent an instrument from reaching the bolt or lever.
Incisions:	The steps in the beard of a key which serve to raise the levers in opening position.
Letter Lock:	See Combination lock.
Lever:	A flat piece of modelled sheet metal of which one or more in a lock must be lifted simultaneously but differently by the various steps of the key to block the dead bolt either in open or locked position.
Lever Belly:	The lower curved edge of the lever which the key touches and slides.
Lever Handle:	An alternative to a knob; to be pressed down for withdrawing a spring bolt.

Lever Handle Lock:	The lock of which the bolt is withdrawn by turning downwards spindle or handle.
Lock Cover:	Is screwed to the lock plate to cover the moving parts and to keep them in place.
Lock Pick (or picking tool):	An instrument of high precision made for the purpose of opening locks when the key is not available. Lock pick is also the name sometimes given to a person who opens the locks in this manner.
Lock Plate:	The plate of a lock in which all the pins, studs and stumps are riveted and to which the front plate is fixed.
Master Key:	The key which passes a number of different locks.
Mortice Locks:	A lock which is inserted into a hole cut in the style of the door.
Interchangeable Lock:	A lock (for safe deposit boxes) interchangeable from one door to another by control of the renter's key.
Key Exchangeable Lock:	A key lock (for safe deposit boxes) operated by several keys but which only opens by turning the key, to the steps of which it has been adjusted.
Key Hole:	The opening by which the key is put into the lock.
Key Way:	The longitudinal cut in a cylinder lock plug to receive the corrugated key.
Knob:	The round or otherwise balanced handle attached to a lock, latch, door or piece of furniture, which may be gripped for turning or pulling.
Knob Latch:	The name used instead of lock for the type having a spring bolt withdrawn by a knob only but no key.
Knob Lock:	A lock having a spring bolt withdrawn by a knob and a dead one moved inwards or outwards by a key.
Knob Handle Lock:	The latch bolt of this type of lock is moved inwards or outwards by a spindle, which can be turned either to right or left.
Latch:	Another name for the type of lock having bevelled spring bolt which is self acting when closing a door.
Latch Bolt:	Another name for spring bolt. It is a bevelled bolt, pushed into the lock case when closing a door.
Night Latch:	A spring bolt lock which can be operated by a key from the outside only and a knob from inside.

Padlock:	A detachable lock of which the swinging shackle passes through a hasp and staple or something similar.
Paracentric Key:	A cylinder lock key with the usual corrugations on one or both sides which reach or nearly reach the centre of the thickness of the bit, thus preventing a flat strip, like a blade of a knife, from entering the keyhole.
Permutation:	Variation in the order, size and form of the incisions or steps in a key bit.
Pin Tumblers:	The name of the pins to provide security in a cylinder lock, of which the upper are known separately as rollers and the lower ones as pins.
Pipe Key:	One with a flat bit and a circular shank drilled to fit on the drill pin.
Renter's Key:	A key supplied by the bank to the renter of a particular safe deposit lock and which moves the lock of the client under the control of the custodian's or bank employee's key.
Rim Lock:	One which is fixed on the inner face of the door style.
Safe Deposit Lock:	One for the door of a locker which a person may rent at a safe deposit. The lock, usually with two keyholes, provides dual control requiring the custodian of the bank to use a key to release the mechanism before the renter, with another key, can withdraw the bolt of the lock and open the door. (See also: Renter's Key).
Servant Key:	A key peculiar to a particular lock in a master keyed suite.
Key Shank:	The part of the key between the bow and the bit.
Single Throw Lock:	One with a bolt which is shot out by one turn of the key only.
Skeleton Key:	A key with a bit which is cut away as much as possible to avoid obstructions in a lock, but with enough left on the bit to lift the tumbler and move the bolt.
Slider:	A small sliding flat piece with notches lifted by a key (e.g. the Bramah key) to move the bolt. A slider usually means a sliding lever.

Spindle: The shaft of a knob or handle usually square in section which passes through the follower to enable the handle, when turned, to operate the spring holt.

Stem of the Key or Key Shank: That part of the key between the bow and the bit.

Strike (Strike Plate): The metal plate fixed to a door style into which the spring bolt of a lock and the dead bolt, if there is one, shoots. This plate is provided with a lip on which the spring bolt strikes.

Stump: A pin (square or round) used in the case of a lock to guide the different parts. Sometimes it serves to receive a screw and is used in lever locks to work in the gatings.

Sub Master Key: When locks are divided into two or more different groups or suites the key which passes some of the different locks in one suite is called a sub master key and the key which controls all the suites a grand master key (See also: Master Key).

Talon: The notch or grab in a dead bolt or runner which the key enters to operate the dead bolt or runner.

Tumbler: A part to retain the bolt or provide security in certain locks. This word is mostly used in the U.S.A. for the English word "lever".

Ward: A fixed projection in a lock to prevent a key from entering or turning, unless suitably shaped.

Warded Key: A key where the bit has been cut away or notched to allow wards in the lock to pass through the bit, as it turns to operate the levers or bolt.

Etymology of the Words, Locks and Keys

Lieut. General Pitt-Rivers' book, entitled *"On the Development and Distribution of Primitive Locks and Keys"*, provided some valuable details on this subject which are summarized in this chapter and, where necessary, have been completed and corrected. The Latin word *"Claustrum"* means lock, bolt, bar or shoot; *"Claudere"* signifies - to lock, to close; *"Clausura"* — a lock and a castle.

The Italian word *"Lucchetto"* means a small spring lock, and Latin *"Laqueus"* — a knob, a noose, a snare, a shackle.

Old English *"Lacche"* is latch or spring bolt in modern English. From the Latin verb *"serere"*, meaning — to join, to connect, the word *"sera"* has been derived, which means "a locking bolt". The Italian word *"Serratura"* means lock and so does the Spanish word *"Cerradura"*, the French word *"Serrure"*, the German word *"Schloss"*, also meaning castle. In Low German *"Slot"* (plural *Slöten*) and in Dutch *"Slot"*. "Slot" is the usual

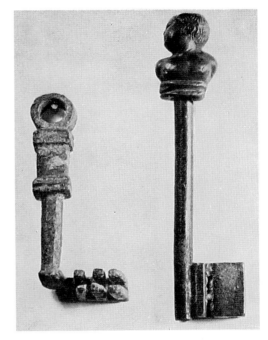

6. Roman Keys. Original in the Lips' collection

word in vulgar English for "latch" and "shoot", as well as in Old Frisian and Low Danish.

The English word "Lock" may be derived from the German word *"Luck"*, *"Loc"* in Anglo-Saxon ; *"Lukke"* in Danish; *"Loca"* in Icelandic; *"Poke"* in Welsh; *"Lock"* in Frisian; *"Loquet"* in French; *"Laas"* in Danish; *"Las"* in Swedish; and *"Glas"* in Irish. *"Klu"* in Greco-Italian means to move, from which *"Sklu"* was derived, in Old-German *"Slut"* means to lock. The Latin word *"Clavis"* represents "key" in English, *"Clavicula"* collar-bone; *"Chiave"* in Italian; *"Llave"* in Spanish; *"Clav"* in Old-Spanish; *"Chave"* in Portuguese; *"Clar"* in vulgar French; and *"Clef"* or *"Clé"* in modern French.

The modern English word "key" has been derived from the Anglo-Saxon word *"Caeg"*, in Old Frisian *"Kay"* and *"Kei"*, in Scottish *"Keg"*, in German *"Schlüssel"*, in Dutch *"Sleutel"* and in Old Dutch *"Sloettell"*, the three last named have probably been derived from *"Sklu"*.

It should be remarked here that the derivations are mostly taken from General Pitt-Rivers' book.

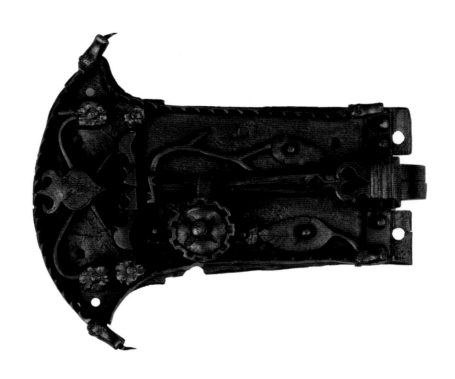

Locks and Keys
Their Origin, History and Use

"Every man I meet is my master in some point in that I expect to learn from him."

- Emerson

CHAPTER I
What is a Lock?

A question easier asked than answered. To begin with it can be said without exaggeration that the history of locks and that of civilization are equally old, but even in the earliest times mankind had to protect themselves and their property against robbery. We all know that envy and egotism led Adam and Eve's children to commit fratricide, and it may be assumed with certainty that stealing and robbery have since persisted throughout the ages.

Man's earliest possessions consisted of provisions, apparel and the like either gained by labour from the soil or by hunting, and records disclose that ancient tribes dug holes in the ground, blocked by heavy stones, or used hollowed out tree trunks, caves and grottos as "strongrooms" to protect their valuable property against the avarice of their fellowmen. Fig. 7 will give the reader an idea of these methods as applied in those days and most likely still applied by certain tribes.

The rude hut, primitive dwelling though it was, marked considerable progress. It was "locked" by means of a detachable door or wall hinged on cords or ropes made of rush or fibre, which served to lock the door or wall.

Cords and ropes appear to have been in use for many years in several countries to lock doors, walls etc. and thus to keep objects out of reach. Indeed they were known as the only security device. It may interest the reader to know that archaeologists in their records of this period refer to the so-called Gordian Knot.

The legend goes that the Gordian Knot could be untied only by one man. Gordius, King of Phrygia, consecrated his chariot to Zeus as a thanks offering, and it became a belief that the knot which bound the yoke of the cart to the shaft could be untied only by the man who would conquer Asia. When Alexander the Great (334 B.C.) arrived at Gordium*, he attempted to undo the knot, and failing to do so with his fingers, cut it with his sword. Thus "to cut the Gordian knot" to-day describes a bold, decisive action effective when milder measures fail. Cord and ropes gradually gave way to wooden doors, locked by means of a wooden bar balanced in the centre of the door, with the bar ends fitting into sockets sunk into the door style. It is curious to find that this system of locking is still applied at present for the double doors of sheds on farms.

*GORDIUM, an ancient city of Phrygia on the road from Pessinus to Ancyra, and not far from the Sangarius. Its site lies opposite the village of Pebi, a little north of the point where the Constantinople-Angora railway crosses the Sangarius. According to the legend, Gordium was founded by Gordius, a Phrygian peasant who had been called to the throne by his countrymen in obedience to an oracle of Zeus commanding them to select the first person that rode up to the Temple of the God in a wagon. The King afterwards dedicated his car to the God. Another oracle declared that whoever succeeded in untying the strangely entwined knot of cornel bark which bound the yoke to the pole should reign over all Asia. Alexander the Great, according to the story, cut the knot by a stroke of his sword. Gordium was captured and destroyed by the Gauls soon after 189 B.C. and disappeared from history. In imperial times only a small village existed on the site, (extract from British Encyclopaedia - Vol. X page 542).

7..... blocked by heavy stones.   *8. Slaves carried the chests....*

Both these turning bars and the subsequent sliding bolts could be operated from the inside only, which inconvenience was finally overcome by the invention of the first real lock.

Although locks, keys and bolts are referred to in the annals of any country, and museums exhibit many specimens, the name of the inventor still remains a mystery, nor shall we ever know when and where the invention was made.

However, the fact remains that the Egyptians, the Romans, the Greeks and other peoples used special vaults with heavy doors in which to deposit the valuables of the wealthy, in a similar manner to that used nowadays in the well known safe deposits and strong rooms of our banks. As a protection against fire, these doors and chests were made of extremely hard wood and in addition were mounted with bronze or iron strips. As shown in the illustration Fig. 8 slaves carried the chests into or out of the vault and clay or wax seals, bearing recognition signs, were affixed to the chests by the "depositors". In most cases the main doors of these primitive "Safe deposits" called *"Horea"* were already fitted with locks.

Maximum security, however, was obtained by the guardians, whose head was the Horerarius key bearer in sign of his dignity and high function.

It is worthwhile mentioning the curious way in which security was obtained in India and the system which has been in use in that country up to the present day.

The "guardian angels" of the valuable property of the Emperor of Annam are a number of crocodiles (Fig. 9). The covers of large hollowed out hard wood blocks and tree trunks containing valuables were sealed with wax and submerged into large pools surrounded by solid walls within the inner courts of the palace. Huge crocodiles – kept on a severe diet – protect these pools, so that nobody could get near the blocks or trunks without risking his life.

9. "Guardian angels".

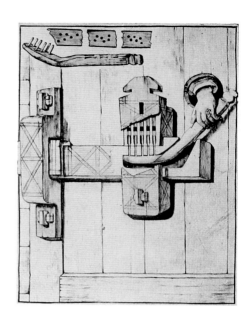
10. Egyptian lock.

Legitimate approach to the treasure is only possible by drugging or killing the animals first.

Until now we have not succeeded in tracing with any accuracy when locks were invented and for how many years they have been practically applied, but the Holy Scripture reveals that locks and keys were known long before the birth of Christ. The classics of that time refer to these devices and in the song of Solomon, chapter V, verse 5, we find the following passage: *"I rose up to open to my beloved, and my hands dripped with myrrh and my fingers with essence of myrrh upon the handles of the bolts"*. References to locks and bolts are also found in Nehemia, chapter III, verse 3 (445 B.C.) and in Judges, chapter III, verses 23 and 25. Nehemia, chapter III, verse 3 reads: *"And the fish gate did the sons of Hassenaah build; they laid the beams thereof and set up the doors thereof, the bolts thereof, and the bars thereof"*. Judges, chapter III, verses 23 and 25: *"Then Ehud went forth into the porch and shut the doors of the parlour upon him, and locked them"* …… *"And behold he opened not the doors of the parlour; therefore they took the key, and opened them…."*. In their encyclopaedias, Rees, Herberts and other writers mention these types of Egyptian locks (Fig. 10). However, the oldest specimen ever found was a wooden lock as described by Mr. Bonomi in his work on Egypt. It was used to secure a single hinged gate in one of the apartments in the palaces of Khorsabad. He writes: *"At the far end of the hall, just behind the first columns, there was a strong single hinged gate, which was closed by means of a heavy wooden lock of a type still used for the town gates and for granary doors. The key for this lock was so big and so heavy that a full grown man could hardly carry it alone. The key could be inserted in a bolt the front part of which entered a large square hole in the wall"*. No doubt it was this type of key the prophet alluded to when he quoted: *"And the key of the House of David will I lay upon his shoulder"* (Isaiah XXII, verse 22). In Eastern countries even nowadays such keys are usually carried on the shoulder.

Equally remarkable is the fact that the name "*Muftah*", meaning key, which is a biblical word, is still commonly used in the East. Such keys may vary in overall length from 10 to 20 inches, according to the sizes of the locks. As already stated, they are carried in a bunch on the shoulder and normally the number of keys per bunch corresponds to the number of granaries controlled by the man. The keys differ in shape, some are straight whereas others are of a bent construction, mostly made of hard wood, although steel ones are also found.

However primitive their design may be, the function of these Egyptian locks is ingenious indeed, and since this 2000 to 3000 year old key has been the forerunner of one of our present day most popular locks, it is worthwhile to explain and illustrate its function and design.

The lock shown in Fig. 11 consists of two main sections of which section A, the lock case, is mounted on the inside of the door by means of strong pegs, while section B represents the bar or bolt which slides in the lock in a horizontal place, and C the corresponding key. The key for this type of lock passes through an oval shaped opening near the top right corner of the lock, as shown in Fig. 10. This opening is cut to allow a hand and the key to pass through. The key is then inserted into the hole of the locking bolt B and then lifted.

The pegs on the key end are set so that a straight line is secured between the lock case and the top end of the shooting bolt, which can now be withdrawn to unlock the door. When pulling or pushing the key the bolt can be moved from right to left and vice versa; when locking the door the left end of the bolt shoots into a staple.

When withdrawing the key after locking the door the pegs in the lock case drop by their own gravity into the corresponding holes in the sliding bolt, which prevents displacement of the latter. This type of lock allows a considerable variety of key permutations, as the pegs can take innumerable positions on the flat end of the key whilst the sizes of the pegs themselves can also be varied.

11.

Arab carrying keys on his shoulder.

Egyptian lock dismantled.

Original in the Lips' collection.

12. Roman lock.

13. Temple of Karnak.

14. A Moroccan lock.

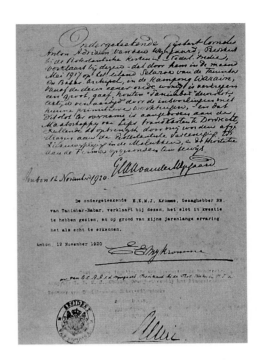

15. Testimonial given by Dr. v.d. Wijngaard of Ambon.

The use of locks and keys in ancient times is further evidenced by the carvings of a sickle shaped key found in a bas relief carved in the columns of the Great Temple of Karnak, the building of which was started over 2000 B.C. It appears that since that time the design and construction of these types of locks and keys have undergone very little change, although sometimes the key was of the straight type, and sometimes sickle shaped. Similar types of keys were found in the tombs of Luxor, very fine specimens indeed, with iron pegs, and carved ivory handles, often inlaid with gold or silver. This type of lock, described here, reminds us of the wooden locks of Persia and Morocco, the principal difference being that they have a separate bolt and a separate key each fitting into their own slide opening.

In addition these locks have square sliding blocks and their keys are provided with corresponding square notches to operate and to set these blocks.

Strangely enough this type of key is also found in the Dutch East Indies in the Tanimbar Islands (West of New Guinea). They are used for locking dwelling houses and fishermen's huts and are used also in the Faroe Islands (North of Scotland).

A written statement received by the writer in 1920 from an Ambon minister together with a testimonial given by a Dutch official confirms this rarity. A photograph enclosed with the statement shows a native of Tanimbar Island demonstrating the lock. From another source confirmation was received that the Tanimbar lock type has also been found in the Moluccas (the Dutch East Indies). A specimen was brought to the Netherlands by Mr. d'Angremond of Rijswijk in 1912 and is now in the writer's collection.

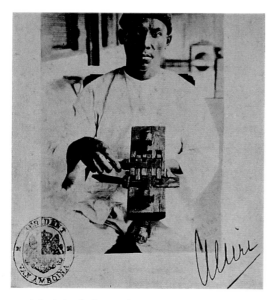
16. *A lock from the Tanimbar Island (Indonesia).*

17. *Tanimbar Island lock with withdrawn key.*

Although its design is definitely reminiscent of types from ancient times, this particular lock was brand new. Even though we cannot speak of an industry, new locks of ancient design seem to have been made throughout the generations. Wooden locks with sliding blocks and operated by wooden keys have also been found in the Neth. West Indies; they are called Maroon locks, because they are mainly used by the Maroons to secure and safeguard their primitive dwellings. These locks are shown in the illustrations in Figs. 14, 16 and 17. The Rev. Father Jac. Mois presented one of these locks to the writer, whilst the other was received from Don José Garcia-Monge y de Vera. I am greatly indebted to this gentleman for his valuable contributions to the Lips' collection.

Fig. 19 shows a Hut lock from the Belgian Congo, which functions exactly according to the principle of the already described Tanimbar island lock and similar locks: Faroe Islands, Bush negro cabin and such like. The original principle was probably introduced in various parts of the world by slaves who transmitted this locking device.

It is made entirely from hardwood and on the top a crocodile has been carved, as an ornament.

The illustration in Fig. 20 represents the Maroon lock from Surinam (1 & 2), the Tanimbar Island lock (3), the lock from the Faroe Islands (4), and the Egyptian lock with sickle shaped key (5).

The Faroe Island lock is considerably smaller than the other types mentioned. However, they all work on practically the same principle, which at the same time implies that these peoples must have had contact with each other.

On a trip through Limburg, a province in the South of the Netherlands, Mr. Kloppenburg, scientific assistant at the Rijksmuseum van Volkskunde "Het Nederlandsch Open

18. Tanimbar Island lock with key inserted in the lock. Original in the Lips' collection.

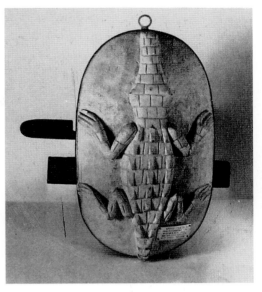

19. Buta-lock with crocodile (Belgian Congo).

-luchtmuseum" discovered a wooden lock fixed to a barn door which reminded him of the Egyptian type of lock and a close resemblance to the locks of Tanimbar, Surinam and the Faroe Islands. The illustration in Fig. 20 - no. 3 represents a specimen of this lock as used on a barn door of a farm near Terstraten (North Limburg). It should be added that this type of lock is still found in several places in Limburg, and also in Germany, France and probably in Belgium as well. For those interested more particulars are to be found in a report on a lecture on the subject "*Schlösser mit Fallriegeler*" by Felix von Luschan, published in the "*Zeitschrift für Etymologie*", in the elaborations about wooden locks by Richard Andrée in the "*Braunsweiger Volkskunde*" and in "*Bäuerliches Hauswerben in Niedersachsen*" by W. Bomann.

Mr. Luschan's lecture is very comprehensive and instructive, although the writer does not agree with him on all points. Even though it is beyond the scope of this book to argue the point, it is worthwhile quoting some lines from it.

In many parts of the Mohammedan Orient wooden door locks are still in use of which the bolt is held in the locked position by wooden pegs, moving in a vertical direction and falling by their own gravity. To release the slide bolt, the pegs are lifted by a wooden key having the corresponding notches. Also in different parts of Germany, Pomerania, East Prussia, Hesse-Darmstad, Alsace and in the Black Forest this kind of lock is still found. There is a large collection of these keys from Hanover in the '"Vaterländische Museum" in Celle, and this collection has been extensively discussed in the treatise: Die Schrift von Billenstein "*Die Holzbauten und Holzgeräte der Letten Mittau 1907*".

This type of wooden lock is also found in Salzburg and Karnten, although they are gradually giving way to modern iron locks. However, wooden locks are preferred in

20.

1 and 2. Lock of a Marcon hut (Surinam).

3. Tanimbar Island lock.

4. A lock from the Faroe Islands.

5. Egyptian lock witch sickle shaped key.

the damp regions where iron locks are apt to get rusty, and for this reason the doors of the shelter huts in the Alps and of the Waldmühlen are fitted with wooden locks. Wooden locks have been found throughout Equatorial Africa as is evidenced by the specimens of different design exhibited in museums.

Von Luschan acknowledges that the wooden lock still actually used in Egypt and called (by him) the Assuan lock is definitely the oldest type known but von Luschan contradicts himself when he discusses the Babylonian type. In the writer's opinion he mixes up different classes of locks, for the design of the Assuan lock differs considerably from the type based on the vertical pegs and is also totally different from the Babylonian lock system. Most interesting details are given about the Roman lock in the Provinzial Museum in Hanover. In his treatise he shows a reproduction of a bronze bolt of which a specimen from Southern Italy is also to be found in the writer's collection.

I think these locks were operated by a bitted or warded key of which the wards and notches correspond to the holes in the bolt, these holes in their turn fitted the vertical pegs. On reconstructing these locks I join him in saying: *Plusque ça change, plusque c'est la meme chose*. Richard Andrée concentrates on the wooden lock mounted on the doors of the Saxon House and tells us about the use of locks of the vertical wooden peg construction in the town of Brunswick. A specimen of this lock is now in the collection of the Municipal Museum of Brunswick. According to him wooden locks were also used in the Carpathians in Romania and Russia as well as throughout the Orient, and they appear to have been introduced into America (Peru and Mexico) by the Spaniards. Likewise Mr. W. Bomann refers to the large collection of wooden locks in the Municipal Museum of Celle, and even to a reproduction of a special type of wooden key, fitted to folding doors of which the bolt is secured by engaging one peg or pin with a circular shank, which can be removed by a finger.

Further references to wooden locks, which on the whole fall in with those of the writers named here, may be found in *"Elsasser Holzschlossen"* by R. Forrer, Antiqua 1889; in *"Volkskunde am Berner Bauernhaus"* by Christian Ruby, Basle 1942.

21. Marcon's Cabin in Surinam.

Justius Lipsius (Joest Lips), the well known Antwerp scholar, who lived from 1547 to 1606, was one of the first to mention the use of locks and keys in remote antiquity in his treatises on the second book of Tacitus, where it states that miniature keys were already used by the Greeks and the Romans. These keys were fastened to rings worn on the finger. Part of the ring had a flattened section into which some kind of signet had been engraved for sealing purposes. Specimens of the Roman type are found in British Museums. This type of key is also included in the Lips' collection at Dordrecht and in John Mosman's collection in one of the halls of the General Society of Mechanics and Tradesmen of the City of New York. Most authors are unanimously of the opinion that the basic design of these keys was of Egyptian origin. It is an established fact, therefore, that the modern well known cylinder locks had their cradle in Egypt, and it is worthwhile mentioning here that it was Linus Yale Jr., son of the founder of the world famous Yale and Towne Manufacturing Company of Stamford (Conn.) U.S.A. who in 1857, guided by the basic construction of these Egyptian locks, designed this first cylinder lock and took out a patent for it.

This marked a revolution in the subsequent manufacture of security locks. These security locks will be fully discussed and analysed in a separate chapter.

Linus Yale Jr. never entered into a partnership with his father. He was born in 1821 and started his career as an artist. However, he soon followed in his father's footsteps, designed and made several types of intricate locks, mostly for applying to doors of strong rooms and safes. To these locks, which indeed are of an ingenious design and show striking features, Yale gave individual names, like: Yale's infallible Bank Lock, Yale's Magic Bank Lock, Yale's double Treasury Bank Lock and Yale's Monitor Bank Lock. Some of these locks were operated by keys, others were keyless combination locks. Linus Yale Jr. took full advantage of his exceptional talent and expert knowledge, continuing his efforts to improve the miniature locks, particularly the key locks for general and wide spread application. The cylinder lock was one of them; it was put on the market in the period between 1860 and 1868 and inspired by the principles of the old Egyptian pin tumbler mechanism which thus was given a new lease of life. This lock has gained a worldwide reputation and may be said to be the most popular in existence, outnumbering any other lock in use.

Linus Yale Jr. entered into partnership with Henry R. Towne in 1868, and to secure an efficient business administration a corporation was founded and a very modern building was erected in Stamford, Conn.

Following Yale's sudden death on December 25th 1868 the business was continued by Mr. Towne and his staff of 30 to 40 employees, and has since developed and extended into one of the largest factories in the world.

CHAPTER II
Metallurgy

The Ancient Egyptians and Romans were familiar with the processes of Metallurgy; they knew how to work iron, although the working of the softer and more malleable metals like copper, bronze and tin were known long before their day.

Copper, silver, gold and other metals can be smelted and cast into moulds. After that they can be easily worked on account of their high degree of softness being easily hammered, bent, pressed and soldered. Iron on the other hand, which is hard and brittle, resists shaping and must therefore be prepared differently to make it malleable and easier to work.

For the benefit of those readers who would like more information there follow details of the properties of these metals since they have played such an important part in the manufacture of locks and keys throughout the ages. Far fetched theories of metallurgy however are avoided as controversial.

Pig iron should be divided into two groups according to its properties, the malleable and the non malleable groups. Pig iron is derived from the blast furnace process and is again divided into dark pig iron and white pig iron. By intense heating and by adding oxygen the metal is reduced and then purified, i.e. all impurities like carbon, silicon, sulphur, phosphor and the like are removed by oxidation.

By adding iron stone, magnet iron stone, manganite etc. to the pig iron in the blast furnace the different mixtures predetermine the properties of the final material according to the purpose for which the metal is required. The metal is then processed in the rolling mills or in the drop forging works to a final product.

After these processes pig iron is still unsuitable for many purposes owing to its brittleness. Grey cast iron is mainly used for castings and white cast iron is processed in rolling mills. The names grey and white cast iron are derived from the colour at the breaking points.

Puddling was the oldest method of melting and purifying cast iron and it consisted in melting the pig iron with iron ore in a reverberatory furnace, in which the reduced metal was continuously stirred (puddled) and the impurities separated. During this process air is forced through the mass of liquid pig iron to speed up oxidation of the impurities.

Slag formed by the oxidation of carbon, phosphorus and silicon was continuously removed. In this manner a fluid mass was obtained, which could be moulded into any shape by hammering or drop forging. This non brittle product of excellent quality is called malleable iron.

However, the process of puddling was improved upon by the invention of the Bessemer Converter, introduced by Bessemer in 1856, and in fact this more modern and rational method entirely replaced the old process in later years. The Bessemer Converter converts cast iron directly into steel. In this process the air, by which the impurities are oxidized, is blown through the mass of liquid pig iron under high pressure and the heat given out during the oxidation of the impurities is sufficient to raise the temperature to such an

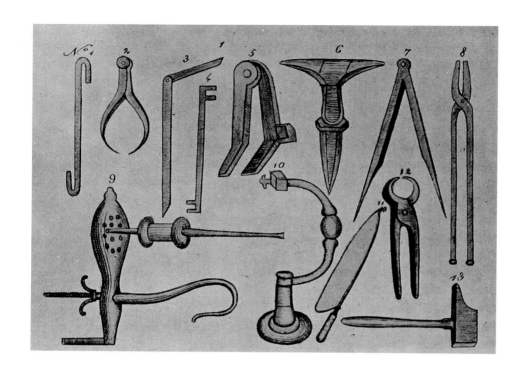

22. Tools of a master locksmith of about 1800. From "Le Parfait Serrurier" by Louis Berthaux, Paris (1828).

extent as to keep the almost pure iron liquid. This modern process is capable of a much higher output.

Although Sidney G. Thomas and Percy C. Gilchrist improved upon the efficiency of the Bessemer Converter, which played an important part in the development of modern industry, nevertheless the Bessemer Converter has been superseded and has given way to methods by which still better quality is obtained, the most important of these being the Bessemer Thomas and the Siemens-Martin open hearth processes, which consist in melting pig iron with various amounts of steel scrap at a fairly high temperature. Crucible steel, the highest quality of ordinary steel, is made by melting pure wrought iron with carbon in crucibles. It was the carbon content of the metal that determined its name, i.e. steel or iron, a discrimination no longer applied nowadays.

In the olden days a locksmith received his blocks of metal from the reverberation furnace and had to re-heat them on a charcoal fire in order to shape them according to his needs, and when appraising his handiwork this fact should not be overlooked. In these modern times lock manufacturers are much better off; they send their blue print to the rolling mill and receive parts in accordance with the design. Some illustrations found in an old French book on lock making by Louis Berthaux (Paris 1828) are worth reproducing here in Fig. 22.

Fig. 23 represents an old smithy where bundles of iron strips were heated and welded to solid bars in shapes: round, square, flat, etc. adaptable to the final product.

Fig. 24 shows the different successive stages of manufacture of an iron key from the component parts up to the final key, whilst Fig. 25 illustrates some types of padlocks of the 17th and 18th centuries, of which sub. 1 represents a letter padlock — a keyless combination lock which can be opened when certain predetermined letters on the rotating rings are brought into line with marks on the lock body.

Top sub. 1 (in Fig. 25) shows the padlock with the shackle in closed position whereas

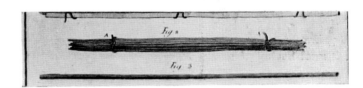

23. *Old smithy where bundles of iron strips were heated and welded to solid bars.*

the bottom sub. 1 shows the shackle when released and how the rotating rings had to be brought into line with the corresponding notches to open the lock. The order of the rotating rings could be changed to suit a different letter combination.
Fig. 25 sub 2, 3 and 4. A series of different models of padlocks with the shackle sliding in the body and operated by a turning key.
Sub 5. The popular spherical padlock with sliding shackle and turning key.
Sub 6. A padlock with sliding key by which the spring loaded bolt is pushed out of the lock case to release the shackle (padlock type derived from the padlocks in use in Eastern countries).
Sub 7. A padlock similar to the one shown in sub 6 but operated by a screwing key and spring loaded rings.
Sub 8. A padlock, the shackle of which is provided with split springs at both ends. This type of padlock was used in practically all European countries.
The keys of many of the ancient Egyptian locks were not made of solid steel, but of sheet iron and the end of the key pin or shank was then provided with a metal bit with ridges, or wards and notches. The illustration given in Fig. 26 shows some specimens of this type of key (Lips' collection), found in Egypt as well as in Arabia.
Some valuable information is given by Mr. George Price and Mr. Francis J. Butter in their excellent books, in which it is mentioned that Roman keys were found in Italy and also in Germany, France, Spain, and even in England.
During the excavations of Pompeii and Herculanum in 1689 numerous keys were found. From their shapes as well as their finish it could be concluded that in these towns iron locks and keys were widely used. Specimens can be found in any private collection as well as in museums in various countries. In the British Museum and the Victoria and Albert Museum in Britain splendid collections are displayed.
In the collection of the Rijksmuseum van Oudheden in Leiden there is a Roman key

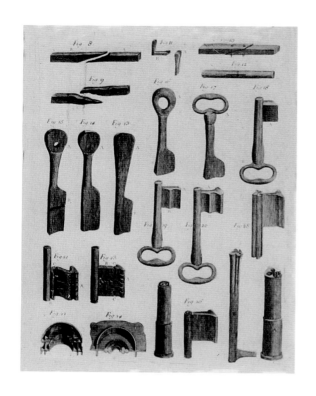

24. Different successive stages of manufacture of an iron key from the component parts up to the final key.

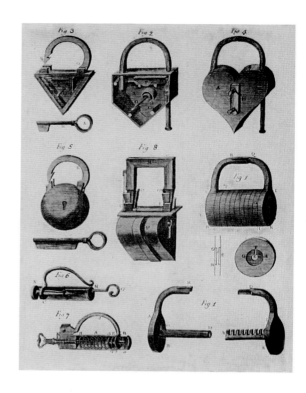

25. Padlocks (Cadenas) of the 17th & 18th centuries.

26. Ancient Egyptian keys, nearly all of them made of sheet iron and bent. Presented by Mr. G. Aivas, Cairo, to the Lips' collection.

which was dredged from the Meuse near Rijswijk in the vicinity of Woudrichem in The Netherlands, which proves that the Romans brought their locks and keys also to this country. The pictures in Fig. 30 give a front and side view of this key, of which the bow or handle was usually made of bronze and of a beautiful design.

The shapes and finish of many of these keys lead to the remarkable conclusion that the Romans were not only familiar with the non turning, that is the sliding, type of key but also with the turning type. Several fine specimens of keys with bows or handles of highly artistic design with carvings, flourishes, etc. are in the writer's collection.

Judging from the design of these keys and the material used — mostly bronze and gold — it may be said that even in those days people set the highest value on fastening and securing doors etc. by means of locks. And not all of these locks were operated by movable parts in the mechanism, by means of wards and obstructions in or around the keyholes. Some were designed to be operated by a ring worn on the finger, so that they could not be forgotten or lost; some had a ring shaped bow to be fixed to the belt. How the mechanism of these locks functioned has so far not been discovered with any accuracy, since the number of locks is considerably lower than the number of keys actually available, and besides this the locks traced have completely rusted, making it impossible to analyse or scrutinize their mechanisms.

The reproductions in Fig. 29 taken from Mr. George Price's treatise (published London 1856) show a number of keys of the above description. Mr. Price adds: *"It is curious to see how many of the Roman types of keys resemble those found in Kent (England) in later years. The construction of the Roman keys already suggested that the mechanism of the corresponding locks contained obstacles, similar to those in the cross-, bar- and warded locks"* (See Fig. 30 and 31).

In his interesting book (1883) Lieutenant-General Pitt Rivers mentions metal objects made by the Ancient Greeks and then continues by saying that Egyptians and Romans also were already conversant with the manufacturing processes of iron and even steel, which by forging they shaped to beautiful objects of art.

He further relates that although the body of the key was mainly composed of wood, iron and bronze parts were often incorporated as well and that by far the majority of keys were made of either of these metals. Specimens of such keys, reproductions of which have been taken from Lieut. General Pitt Rivers' book, were traced in several parts of Europe; the bulk came from Greece, Italy and France, but some also came from England and even from Scandinavia. It is worth noting that the use of the three pegged type of key has been maintained in Sweden and Norway to operate wooden locks on woodcutters' cabins. Some of these keys of Norwegian origin are in the Lips' collection.

In the chapter *"Der Tempelschlussel"* (the Temple Key) Hermann Diels (1897) gives a most interesting picture of the use of keys and their symbolic significance in the days of the Ancient Greeks, and how these keys were carried on the shoulder by the Priestesses, as shown in the reproduction of Fig. 32 and 33. In this connection reference is made to *"Lehrgedicht des Parmenides"*, Berlin (1897), George Reimer and Antike Technik, Leipsic (1914), Teubner.

In the latter part of the 19th century many such temple keys were found in Lusio (Arcadia). Symbolic signs of these keys were also found on vases and pottery as decoration. Their shape reminds us of the human collar bone.

27. Roman ring keys. Lips' collection.

A further evidence of the early use of locks in Greece may be found in the following extract from Homer, which dates from about 700 B.C. : *"Telemachus had spoken to the Freemen for the first time and went home during the night. The young Prince crossed a dark inner court of the palace to go to his sleeping apartments, lighted with a torch by Eurykleia, who unlocked the door for him to enter"*.
And in verse 441 we read:
"Eurykleia came out and gently pulled the silver handle to close the door and secured the bolt by pulling a leather strap".
Latch locks of this kind were operated by a cord or strap, properly twisted to secure or release the bolt.
According to Homer (Odyssey XXI) Penelope took a brass sickle shaped key with an ivory bow to unlock a wardrobe.
In his marginal notes on the fourth book on Odyssey, Eustathius ascribes the invention to the Lacedaemonians. On the other hand, Pliny I credits a certain Theodore of Samos with it.
Aristophanes vividly describes the annoyance of the housewives at the Lacedaemonian locks which, after having been secured by their husbands, prevented the women from entering the larders or store rooms. This type of lock is still found on many street doors in Politiko, Arabia, Huetius, Telemak, Ariston and other places. A collection of ancient Roman and French bronze and iron keys of both the turning and the sliding type are discussed and reproduced in Dechelette's *"Archéologie Celtique – Second Age du Fer"*, in which he says: "Iron locks and keys were already applied in the century preceding the birth of Christ, both north and south of the Alps."
The Greeks had their Temple keys of 25 to 30 cm long-sickle shaped irons with a bow for hanging on a nail or to suspend by a piece of cord. The other end of the shank was bent or hooked and by this some part of the lock mechanism could be operated to release the sliding bolt. Like the Egyptian bolts, these locks were mounted on the inside of the door and the key had to be inserted from the outside through an oval circular hole in the door face to operate the lock.
The use of this type of lock lasted long; even after more modern locks were invented.

28. Iron and bronze Gothic and Merovingian keys. Lips' collection.

30. Ancient bronze Roman keys. Lips' collection.

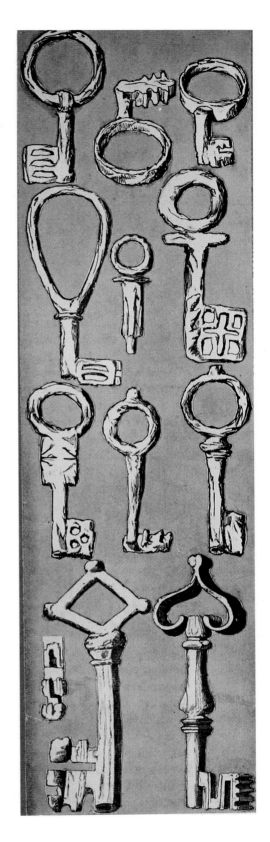

29. Bronze and iron keys from Pompei and Herculanum (George Price).

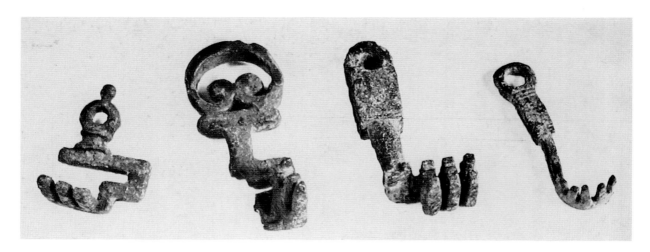

31. The oldest Roman and Greek keys known to us. Lips' Collection.

According to Lieutenant-General Pitt Rivers, the Romans, the peoples north of the Alps, even of Bohemia, Northern France and of the British Isles, in Mount Caburn near Lowes were already familiar with the manufacture and use of locks and keys.

The keys of the Hellen's also remind one of the type of key described here, although wooden keys were generally used as well, which are supposed to be of Egyptian origin and to have been brought to Ionia, to Greece and to Rome.

Finally, Dechelette refers to "*Entraves*" the so-called shackles or fetters and slave chains, operated by padlocks, also used for galley slaves. A specimen, as shown in Fig. 37, is in the Lips' collection and originates from the African Coast.

The chain with which Saint Paul was fettered and led through Rome, by a Roman soldier, has been pictured in National Geographic Magazine Vol. CX No. Six of December 1956 page 157.

This chain is still being kept at Rome and has the same form of links as the slave-chain in Fig. 37.

In England, Germany and France symbolic representations of keys were found in coats of arms and seals as far back even as the 15th and 16th century.

A good example is found in the coat of arms of Admiral Sir C. C. Parr R.N. of Scotland. Judging by the word "Try" (which implies: try to unlock) in one of the reproductions, see Fig. 38, we gather that considerable value was attached to keys as security devices, when designing such coats of arms.

Keys are also symbolized in the coat of arms of The Hon. Charles Kenney (Ireland) in which two coupled or interconnected keys represent the heraldic device: Prudentia et Honor. Although no keys are shown in the coat of arms of Sir Howard Douglas R.N. of Loch Leven, who took part in the siege and bombardment of Flushing in 1810, trust in security by means of locks and keys was expressed by the slogan: "*Lock Sicker*" Lowland Scottish for English "Safe Locking".

The Greek Goddess Hekate is represented by a key, which gave access to the lower world. Many tombstones have the old family coat of arms cut into them.

Leyden, known in the Netherlands as the city of the keys, shows on her shield two crossed keys supported by two lions rampant (Fig. 41).

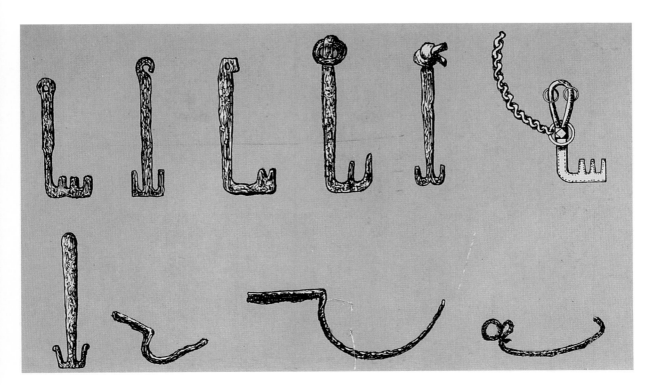

32. From J. Dechelette's: Archeologist Caltique.

33. The Greeks had their Temple keys.

34. The Temple Key from Hermann Diels.

35. Keys Symbolized in Ancient Seals.

36. The Temple Key from Herman Diels.

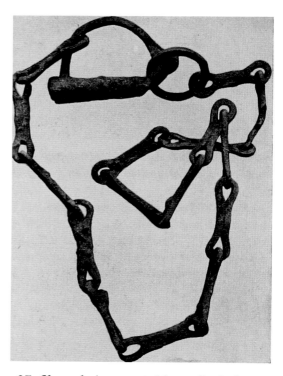

37. Slave chain operated by padlock, from the African coast. Lips' collection.

38. Keys symbolized in Coats of Arms and Seals.

40. Anastasis mosaic in the monastery church of Dafni showing Christ descending in the front court, where He tramples Hades, King of the lower world.

39. Egyptian Sarcophagy-key of King Ptolmis, approx. 300 years B.C. Lips' collection.

Fig. 44 represents the arms of the City of Bremen, the shield exhibits a silver key; however, in Botho's description in *"Cronecken der Sassen"* printed in 1492 by Peter Schoffer in Mainz, a shield with two-crossed keys is discussed.
In the Museum of St. Annenkloster of Lübeck a shield is displayed, the right half of which shows an ancient Roman key.

Leyden

As a Dutchman the writer was, of course, extremely keen to trace the origin of the keys on the escutcheon of Leyden, but no definite sources could be traced, although the archives of the City of Leyden supplied some documents and reproductions which were of great use.
In No. 4, part C, page 47 of *"Corpus Sigillorum Neerlandicorum"* published by Wouter Nijhoff in The Hague, reference is made to the oldest city still known to us. According to documents found the seal was used in 1293 and 1299 and these documents have been at different times in the Public Archives of The Hague and in The National Archives of Hainault. The centre of the seal shows St. Peter with the key in his right hand.
Mr. S. van der Pauw, historian and architect of the early 19th century (Leyden), whose

41. *The coat of arms of the city of Leyden (Holland).*

42. *Arms of the city Geneva.*

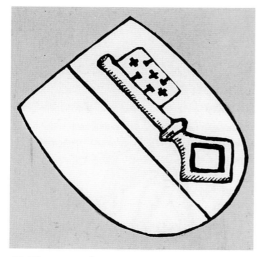

43. *The coat of arms in the Museum of St. Annenkloster of Lübeck.*

44. *The coat of arms of the town of Bremen (Germany).*

investigations provided most reliable information, stated in one of his manuscripts, No. 4, (kept in the archives library), that two further seals of earlier dates were known to him, of which he provided sketches A. and B. in Fig. 45.

In his short notes on "*Lugdunum Batavorum*" (now Leyden) published in 1672, Simon van Leeuwen gives some particulars on the earliest city seals and refers to the one reproduced under D (Fig. 45) which is of a later date than the one under C. Some people concluded that two keys have been adopted to symbolize the actions "locking and unlocking" but van Leeuwen's personal opinion is that the two key symbol in the arms of Leyden is a true imitation of the coat of arms of the Pope of Rome (Fig. 41).

The greatest historian of the City of Keys, Frans van Mieris in his standard work (1770) qualifies the seal as reproduced under D as being of a stately cachet and explains the double key as a symbol of the union of municipal jurisdiction with that of the burg graves of Leyden, thus point authority was expressed.

The coat of arms of the town of Soest in Westphalia dating from 1570 shows a Gothic key.

This shield was affixed over the entrance to the Grandweger-Tor-Bastion (See Fig. 46). A most remarkable shield, showing a key emblem, is that of the fortress town of

Schlüsselburg, north of Leningrad. Naumberg on the Saale, Schlüsselberg in Bavaria, Regensburg, where the Porta Pretoria Romana was excavated, and many other places of interest, all show keys in their coats of arms (Fig. 46).

The town of Le Mans in France for example has a key set between the fortress and the French Lilies as a centre piece (Fig. 46).

Another proof of the great importance of keys as safety, security and reliability devices may be inferred from the use of the name *"Die Sloetell"* (the key) given to a brewery founded in Dordrecht as far back as 1433, the oldest brewery in the country and still in existence. The Deeds of Property dated 11th March 1506 contain the following lines, shown to the writer by the proprietor of 'Die Sloetell':

"Floeris Barthhoutz, for half and Aoffgen Claes, the widow of Alcmade for a quarter share, sell to Jan Gheritz die Coman, the premises on Poortzyde lying between Wouter van Doilkom and Peter Hermansz, which premises are known by the name of "Die Sloetell"

The premises of 31, Groenmarkt, Dordrecht, in the old renaissance stepped gable style, still have a key in the gable stone, although they do not belong to the brewery any longer. Many more examples are available in city arms of seals and shields bearing keys as symbols, among these those of Sloten (3 padlocks), Rhenen, Breukelen (St. Peter and two keys) Abcoude (2 keys), Apeldoorn (eagle and two keys),

Loppersum (key and sword), Terneuzen (swimming lion and two keys), Gilze Ryen (Saint and two keys), Vught (two churches and key) etc.

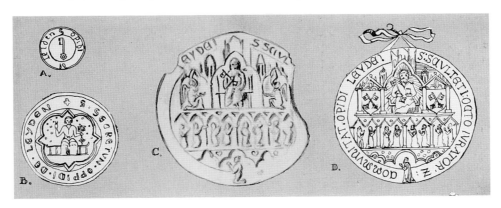

45. Seals used in 1293 and 1299. The centre shows St. Peter with the key in his right hand.

46. Soest   Schlüsselburg   Le Mans   Naumburg
   (Germany)   (Russia)   (France)   (Germany)

Several noble families also show keys in their coats of arms e.g. Schimmelpenninck (two black crossed keys), Klerck (ditto), v.d. Goltz (a golden key and a silver sword crossed) etc. Finally, the Conventual Church of Dafni holds an Anastasius mosaic showing Christ descending in the front court, where He tramples Hades, King of the lower world. In the reproduction of this mosaic keys are shown (Fig. 40).

Key of the Loch Leven Castle

Found in the lake, and supposed to have been the one thrown in by the young Douglas when Mary Queen of Scots made her escape. The key was originally in the possession of William Hamer, Esq., who presented it to Sir Walter Scott, having first had an engraving made of it. The Castle of Loch Leven is situated on an island of about two acres, near the north west extremity of the lake. Queen Mary, when she dismissed Bothwell on Carberry Hill, and joined the insurgents, was carried captive into Edinburgh, and on the following day committed to Loch Leven Castle. On the 25th of March, 1568, she attempted to escape from thence in the disguise of a laundress, but was frustrated. On Monday, May 2nd, 1568, however, while the family were at supper, the boy, Wm. Douglas, secured the keys of the castle, and gave egress to the Queen and her maid from the stronghold; then, locking the gates behind them to prevent pursuit, he placed the fugitives in a boat that lay near at hand, and rowed them to the appointed landing-place on the north side of the lake.

An interesting article about the key of Loch Leven Castle is reproduced with Fig. 47. This key was found in the loch and is said to have been thrown in by the young Douglas when Mary, Queen of Scots made her escape. After her escape pursuit was prevented by locking the gates behind her. In the same way keys were frequently symbolized by the Romans and Greeks. Both the twin faces of Janus and Portinius were represented by keys as a sign of their power to open the gates of the Earth and Sea.

Lieutenant-General Pitt Rivers' book has, in many ways, been a valuable source in the field of locks and keys and on padlocks in particular. Even the ancient Egyptians were familiar with padlocks made of bronze, brass or iron, operated by sliding or turning keys. The Romans, too, knew the principle of turning keys.

The padlocks used by the Arabs, the Persians, in short by all the peoples of the Orient were operated by keys of the sliding type. Padlocks operated by a turning key were rare. A characteristic of the padlock in the Orient is that its outline is frequently that of an animal, a practice also followed in art and architecture of those days (Fig. 48).

In Fig. 49 the reader will find an illustration of a padlock still in use in Far Eastern countries. It consists of two sections, viz. the lock box containing the mechanism, and a separate shackle with a tubular pin at the top and a set of springs in the lower part. When the key is inserted in corresponding keyhole the springs are compressed to an absolutely level position, thus both form a straight line. Now the horizontal springs can be pushed through by the key so that respective spring terminals project from the lock box. The length and shape of the key determine the proper function of the lock and its security. By simply pushing the shackle into the lock box, the springs are expanded, the spring terminals caught by two flanges and by this operation the lock is secured.

Sub 1 in Fig. 56 shows the lock in closed position; sub 2 demonstrates the open position, whilst sub 3 is a sketch of the operating key.
The other sketches (in Fig. 56) show the typical shape of the keyhole and key bit. Padlocks built on this principle were found by the writer in many other countries and the Lips' collection has specimens of English, Spanish, Arabian and other origins (see reproductions in Fig. 50-55). The Lips' collection also comprises a number of padlocks of the screw locking type, which were used in many European countries.

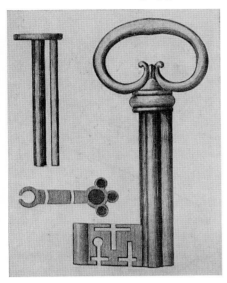

47. *Key of Loch Leven Castle.*

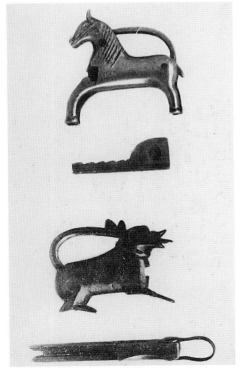

48. *Imitations of different species of animals. Lips' collection.*

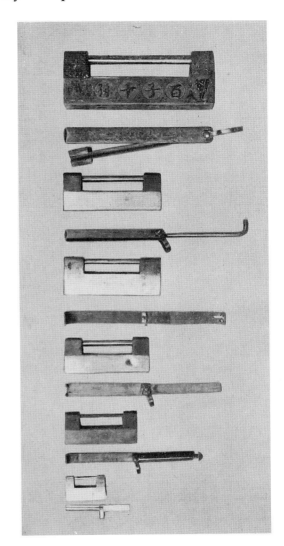

49. *Chinese and Japanese padlocks with sliding keys. Lips' collection.*

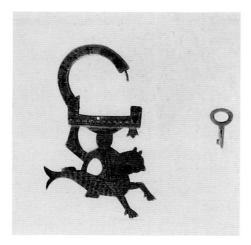

50. Hindu padlock with turning key.

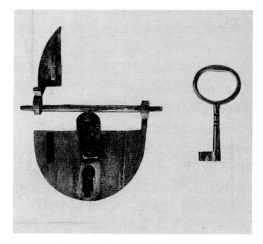

51. Arabian padlock with turning key.

52. Moroccan Vey shaped padlocks with storage accomodation for key. Lips' collection.

53. Japanese padlock with secret flap locking device. Lips' collection.

54. Arabian padlock with screw key. Lips' collection.

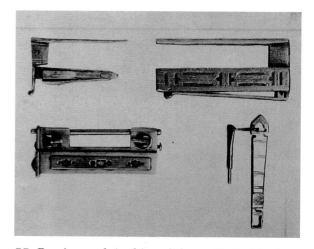

55. Persian and Arabian slider padlock. Lips' collection.

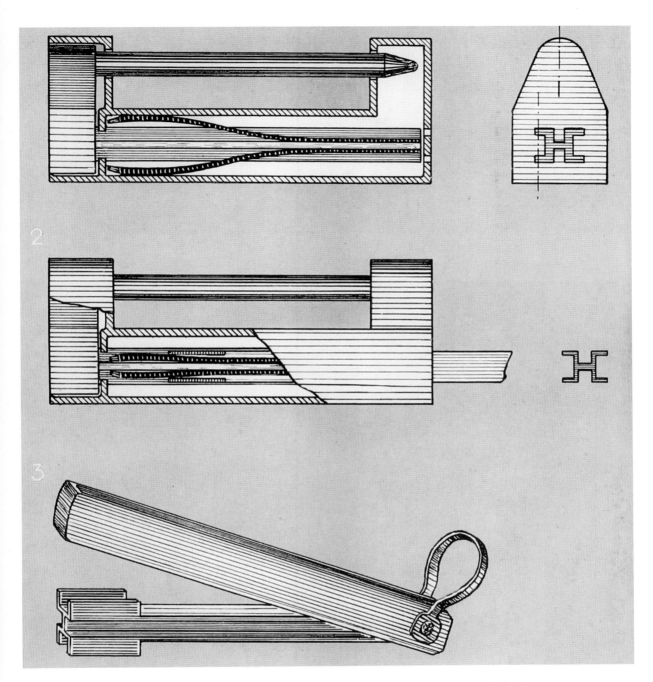

56. 1) Chinese lock in closed position. 2) Lock in open position. 3) The key.

The writer was presented with some padlocks by friends in Scandinavia, as shown in the reproduction in Fig. 57-58. The shackles of these locks, the tub shaped lock box of which contained a set of rotating discs, were released or secured by means of keys, whereas in other countries key shaped padlocks were found of which the key bow served as the shackle, to be opened or secured by another key. A number of the last named types of keys in the Lips' collection were brought from Morocco and Egypt. Mr. A. M. de Jong in his booklet "*Terminologie rörande lås, låsdear och tillbehör*" describes a Swedish padlock called the Polhem lock after its maker Christoffer Polhem, born in 1661, and shows a

reproduction of a lock identical to the one reproduced in Fig. 57.
This Polhem lock is looked upon as the fore runner of the so-called Chubb and Protector locks. Obviously his description of the lock mechanism is exactly the same as that of the Scandinavian locks just mentioned.
From Russia the writer received a splendid extra heavy padlock. The solid bracket-shaped forged lock box also contained a mechanism of rotating discs, which could be set by the different steps in the key bit. This lock dates from the 16th century (see reproduction in Fig. 58)
Elaborate and heavy padlocks (Fig. 62-63) are still regularly used in South European countries, for instance, to secure the town gates, but the application of such locks to secure prison doors and entrances to castles was known long before in Italy, Germany, France and England.

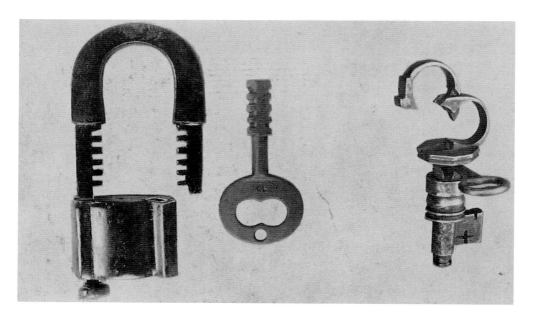

57. Scandinavian padlock from Christoffer Polhem (1661-1851) called Pohlemslas.

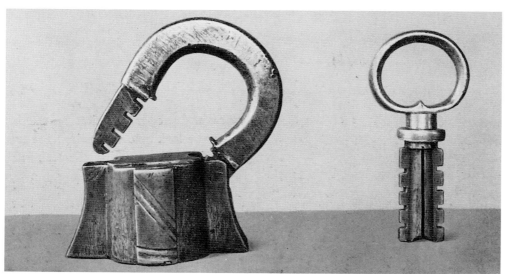

58. Russian padlock (16th century).

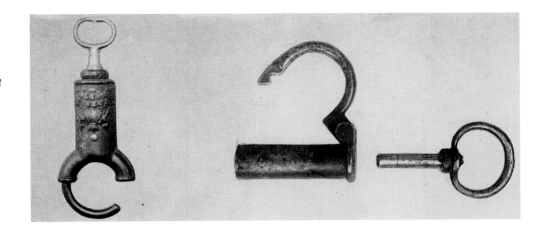

59. Old English padlock. Old Spanish screw key padlock from Burgos.

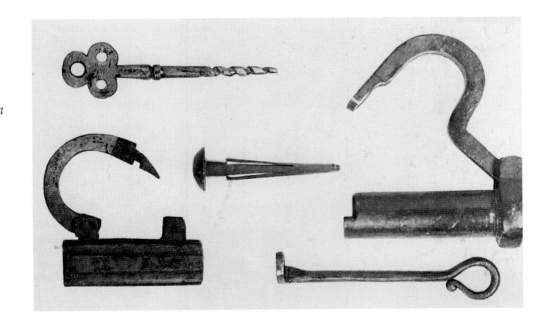

60. Old English screw key padlock. Old Spanish push key padlock from Seville.

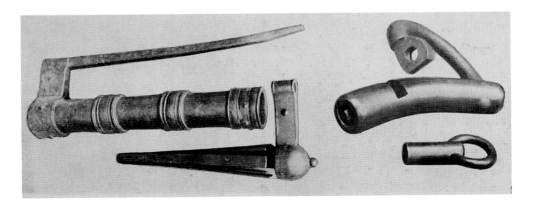

61. Large slider padlock used in the Navy (length about 50 cm), gun shaped, probably English (17th century), English screw key padlock. Lips' collection.

62. A curious specimen of an early Spanish padlock. In locked position. Lips' collection.

63. In open position (Found in the province of Zealand). Lips' collection.

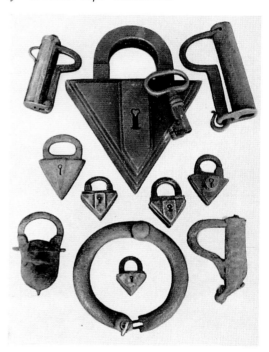

64 and 65. Specimens of European padlocks of the 14th and 15th centuries. Operated by screw, sliding or a turning key. Lips' collection.

A most curious specimen of an early Spanish padlock is reproduced in Fig. 62 and 63 in locked and open position. It was found in the Dutch province of Zealand and most likely left behind by the Spaniards after the occupation of the Netherlands in the 17th century.

Special attention is drawn to the double shackled lock in as far as it has a most peculiar keyhole which can be covered by a secret lid. The pipe keys of both locks fit on a heart shaped drill pin (Fig. 66 and 67). The illustration in Fig. 68 represents an old Dutch padlock with twin shackle and double secret door.

66. Spanish and French 17th century padlocks with twin shackles and double mechanisms. Lips' collection.

67. In open position. Lips' collection.

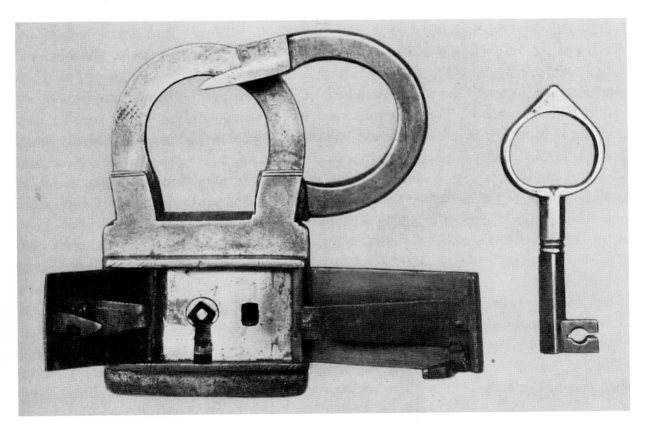

68. Old Dutch padlock with twin shackle and double secret doors. Lips' collection.

69. An ancient and modern keyless combination padlock, together with a modern Lips' cylinder padlock. Lips' collection.

Some further specimens of European padlocks of the 14th and 15th centuries are reproduced in Figs. 64 and 65. They are operated by a screw, sliding or a turning key. Several of these locks are fitted with secret buttons, which needed to be pushed before the key could be inserted.

In Holland and Flanders, too, double shackled padlocks were known, some of which had double covers to the keyhole. Such covers could be opened by pushing some secret button or latch (see Fig. 68). Finally an ancient and a modern keyless combination padlock are shown in Fig. 69, together with a modern Lips' cylinder padlock. The longitudinal section of this cylinder padlock shows the great similarity of this pin tumbler mechanism to the wooden peg mechanism of the ancient Egyptian locks. In fact the locking principles of this modern cylinder padlock are based on those of the ancient Egyptian locks already described in an earlier chapter.

With reference to keyless combination locks, letter locks have been known for many years. The following passage is from Mr. Francis J. Butter's book: *"The keyless letter lock has a number of rings, mostly four or five, which can be rotated individually by hand, until a certain predetermined combination of letters has been reached, after which the lock can be opened. Letter combinations on such locks are either fixed or may be adjusted as required. As soon as the rotating rings are brought into line with marks on the lock body, the shackle is swung on a pivot and thus released"*.

In Beaumont and Fletcher's play *"The Noble Gentleman"* written in 1615, we read: *"with a strange lock that opens on A.M.E.N."*

This lock, designed by Mr. Carden, had a fixed letter combination, which was perfected later on by Mr. M. Requier, Administrator of the Musée d'Artillerie of Paris, in such a way that letter combinations could be altered according to the user's wishes.

Locks and Keys in the Middle Ages

Comparing the oldest types of locks and keys known to us with those used up to the period of the Middle Ages, it must be said that since the time of the ancient Romans and Greeks these safety devices have undergone hardly any changes in design, construction or finish. In fact it was not until the early part of the mediaeval period that the social developments in Germany had their influence also on the technique of locks and keys, and it is for this reason the Middle Ages may rightly be called the second important period of the technical development of locks and keys.

In this period the Germans had grown into a keen and active agricultural nation whose principal means of subsistence lay in the yield from their fields and forests. Articles, serving as decoration or ornaments, like clothes, jewels, iron, gold, silver were "imported" by foreigners, such as the Romans, the Celts and the Jews. Last but not least, their wines were an important commodity imported by these people. All these goods were either paid for in money or bartered against German agricultural products, furs, goose quills, ham, bacon and soap in particular. In the 15th century (1411) Charles IV created the title of Master Locksmith in Germany, which title was conferred on craftsmen having undergone severe tests in ornamental art, for irreproachable workmanship in the construction of locks, hinges, fittings and keys of decorative and artistic design. In the early part of the 16th century, lock manufacture had advanced to such an extent that locks with open covers could be introduced. This type of lock did not only show the component parts of the mechanism and their functions, but at the same time the highly artistic design of the lock case and the method of assembly of the mechanism bore witness to great craftsmanship.

The expert smith devoted himself not only to the manufacture of hinges and ornamental iron work for doors but also to the production of locks and keys, and it is evident that this all-round sense of beauty was extended to and reflected in the artistic design of his locks and keys; in fact it was the smith's strong leaning towards artistic design that proved detrimental to the security and practical size and shape of the products.

The German and French locksmiths of the Roman period and the Middle Ages were indeed famous for their achievements in producing unique specimens of expert smith's work.

A good number of locks and keys of those days have been collected by Museums and may be considered lasting witnesses of the exquisite taste and high craftsmanship of the period. The numerous reproductions of keys of those days in Figs. 70, 71 and 72 testify to this, and the reader will not be long in finding out for himself that many specimens are real masterpieces of smith's work and fine handicraft, for which only very simple, even primitive, tools were available then.

The Roman, as well as the Merovingian and Karolingian keys were mostly made of bronze. The reproduction in Fig. 70 shows an interesting collection of Arabian, Roman and Frankish keys of the sliding type, some made of bronze, some of iron inlaid with brass. In Figs. 73 and 74 a number of forged iron keys are shown in chronological order. The French, German, Italian and Spanish locksmiths were unparalleled in their art of making locks and keys of highly ornamental design. They also made the so-called Chamberlain key worn as a distinction conferred on court dignitaries. This kind of key was also worn by valets and butlers in France, who served the wines. As several specimens of these keys in the Lips' collection show, some have been made of copper with gilded ornamental bows, some have forged iron shanks with copper or bronze bows richly decorated, carved and polished.

Anyone who is in a position to visit the collection of Chamberlain keys bequeathed by Mr. Octavius Morgan to the British Museum in 1888 will still better appreciate the considerable development in the technique of locks from the 15th to 19th centuries. This collection comprises specimens from Spain, Portugal, Germany and Denmark and the illustrations (See Fig. 75-82) can only partly convey to the reader the great craftsmanship and ornamentation.

In the same way the Musée du Louvre, the Musée de Cluny in Paris and the Musée de Ferronnerie in Rouen each hold a marvellous collection, of which the large group *"Le Secq des Tournelles"* is outstanding and, therefore, worth mentioning separately. Illustrations of some keys of this group are found in Fig. 78 of which the four smaller keys in the centre with gilded ornamental bows were used for furniture and smaller-sized chests.

One key of Fig. 78 with its very thin double pin and two other double pin keys in the left and right bottom corners are noteworthy for their two key bits. The sliding part of these peculiar types of keys is in the middle of the double pin and can be shifted along the pin to cover either bit. The key in the bottom centre with the very thin pin has a splendid bow with astronomical symbols.

70. *Arabian, Roman and Frankish keys, mostly made of bronze. Lips' collection.*

71. A number of bronze keys in chronological order. Lips' collection.

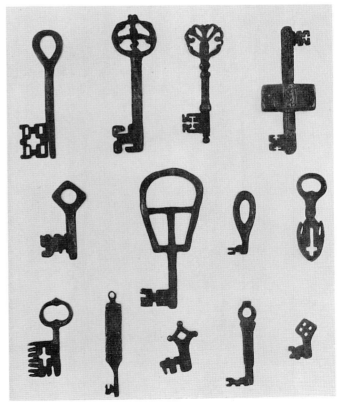

72. Roman, Frankish, Merovingian and Karolingian keys. Lips' collection.

73. a) Barrel keys with fashioned shanks and some specimens of sliding keys. b) chest and trunk keys. c) Cupboard and furniture keys d), e) and f) Keys with ornamental bows dating from various periods. Lips' collection.

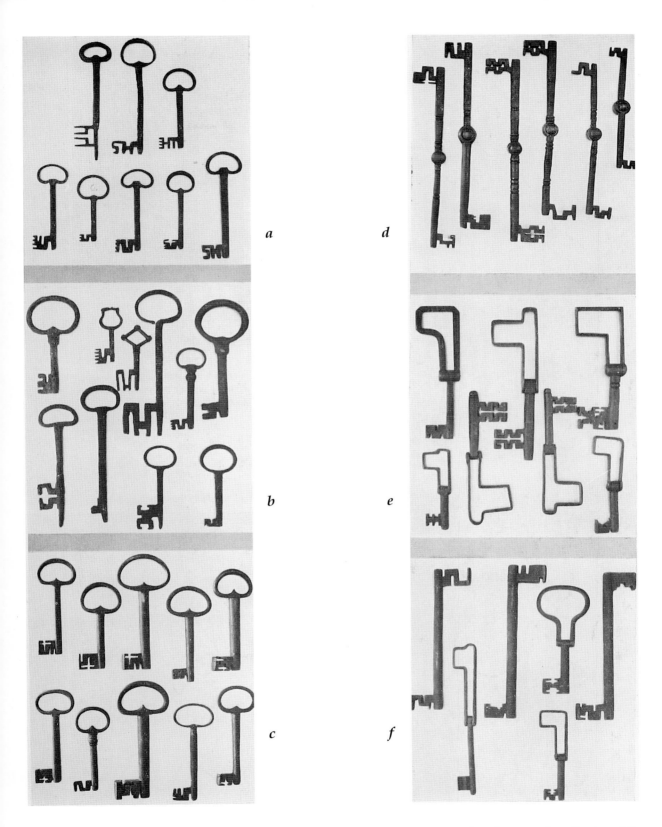

74. a), b) and c) keys dating from various periods, mostly from the 16th and 17th centuries. d), e) and f) folding keys and so-called twin keys of the 18th century.
Lips' collection.

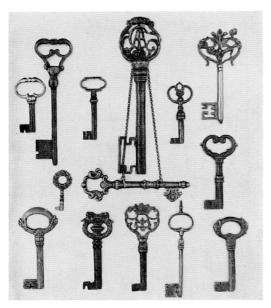

75. Keys of highly ornamental design. Lips' collection.

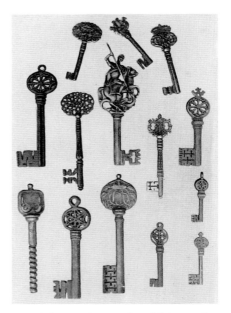

76. Masterpieces of smith's work. Lips' collection.

77. Group of ornamental keys of the Lips' collection.

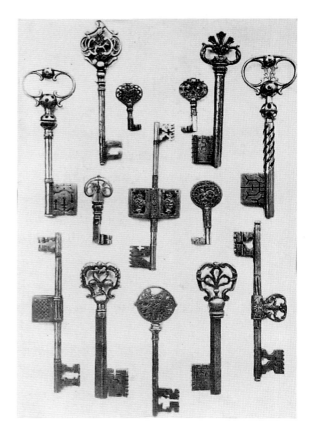

78. Some groups of keys from the collection "Le Secq des Tournelles"

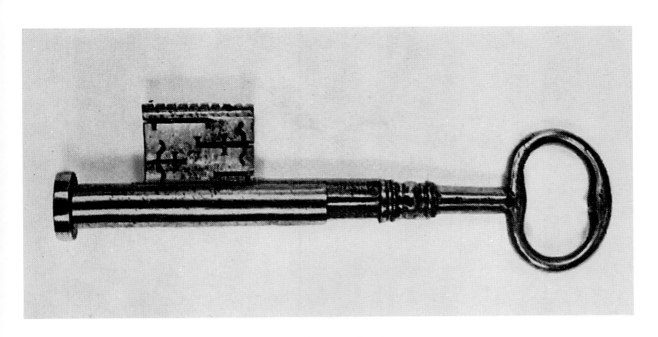

79. Renaissance pipe key.

80. Musèe de Cluny: Specimens of French Art of the period from the 15th to the 18th centuries.

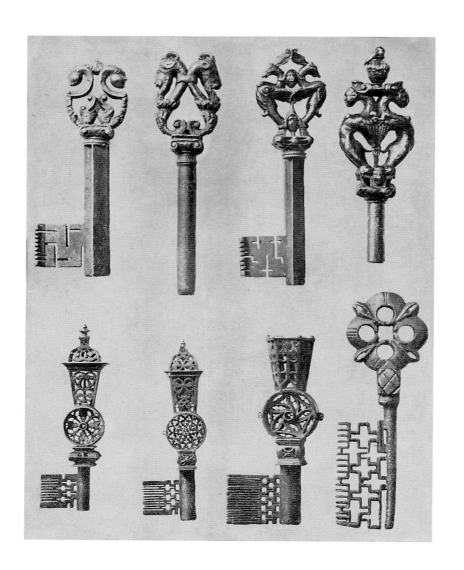

81. Museo Nazionale (Bargello National Museum), Florence (Italy). French keys from the Carrand Collection.

82. Städtisches Museum, Nürnberg. Iron keys with copper ornamental bows (German art).

CHAPTER III
The First Metal Locks

In the preceding chapters the study of the history of locks and keys will have convinced the reader that from time immemorial people have been familiar with their use and manufacture, and since the art of metal working developed in the different countries the standard of lock technique gradually improved with regard to design, construction, process of manufacture and security.

The earliest mention of metal locks was in 870—900 A.D. and the first metal locks are supposed to have been made in England during the reign of King Alfred, who is said to have been a great patron of the art of lockmaking. It may be added here that since those days English locksmiths have lived up to their reputation as makers of high-grade locks.

An important lock industry is located near Wolverhampton and neighbouring districts. According to Mr. Butter, English made locks enjoyed an excellent reputation as regards construction, security and ornamentation even as far back as the 12th century. Judging from a bill for smith's work made out in 1709 by an English locksmith, who for the amount of £2/4/4 went to a lot of trouble to supply lock parts and to do a repair and maintenance job, wages were certainly not very high considering the standard in those days.

The original bill, reproduced in Fig. 83, is in the writer's collection. Practically every museum in England holds a collection of some size, containing specimens of locks and keys found in various parts of the country. It has already been remarked that, generally speaking, the number of ancient keys preserved in various collections exceeds the number of ancient locks.

Guided by the shapes and designs of a large number of these ancient keys, efforts have been made to reconstruct the exact way in which these keys operated the mechanisms of the corresponding locks, but these efforts

83. A locksmith's bill (1709) in the Lips' collection.

have been only partly successful, since any conclusions as to the function of these keys are merely the result of guess work. However, it may be said that the basic idea in designing keys — no matter whether they were sliding, lifting, or turning keys — was to provide the keys with most complicated key bits and at the same time to incorporate into the lock mechanisms all kinds of obstructions fitted in or around the keyhole.

All these devices practically excluded illegitimate opening of such locks.

As already mentioned, the period of the Middle Ages is rightly reputed as a period of great progress in producing locks and keys of highly artistic design. The Germans, the French and the Italians particularly showed themselves masters of their craft and even nowadays their achievements in this field command our admiration. Love of their work inspired them to manifestations of ornamental art, which bear witness to the highly cultivated artistic taste and skill of both designers and mechanics.

However, our admiration is not restricted to the fine outward appearance of the locks, but also to the finish of the lock mechanisms proper, all specimens of craftsmanship despite the comparatively primitive tools of those days.

For literature on this subject the reader is referred to the works of Andreas Dillinger of Vienna, Mathurin Jousse of Paris (1627), Hermann Diels of Berlin, Louis Berthaux of Paris, to mention a few, which will disclose to those interested the high standard of craftsmanship of those days. The number of illustrations contained in these works also reveal the great knowledge of the art of working with different classes of metals and their alloys.

It is with the special consent of the respective owners that the writer is enabled to give a complete survey of their collections. The names of the collectors have been added to the reproductions of their keys and the author wishes to express his gratitude for their kind co-operation.

Frans Engels' Collection - Antwerp, Belgium

As an enthusiastic key collector and a great lover of art, Mr. Frans Engels has gradually built up a most attractive collection. For the greater part these keys — the locks are missing — have been dredged from the river Scheldt, near Antwerp.

Some years ago a booklet entitled *"The Key Collection of Frans Engels, Antwerp"* was published and some pages are shown in Figs. 84 and 85. The keys have been systematically classified in chronological order. Indeed this collection is worthy of mention, particularly as Mr. Frans Engels has never had any connection with the lock trade.

Also in the museum "Het Vleeshuis" on the Scheldt Quay in Antwerp a most remarkable collection of locks and other smithy work can be seen.

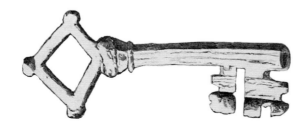

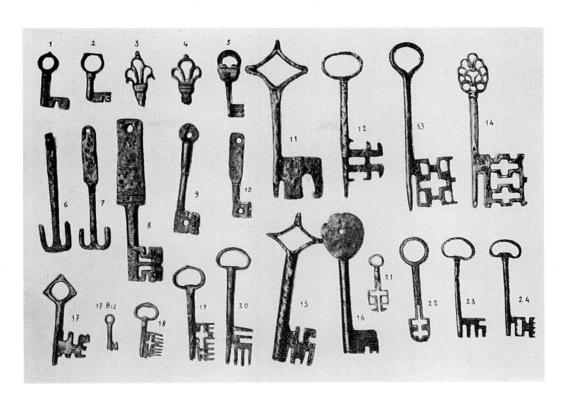

84. Collection Frans Engels, Antwerp. Belgium.

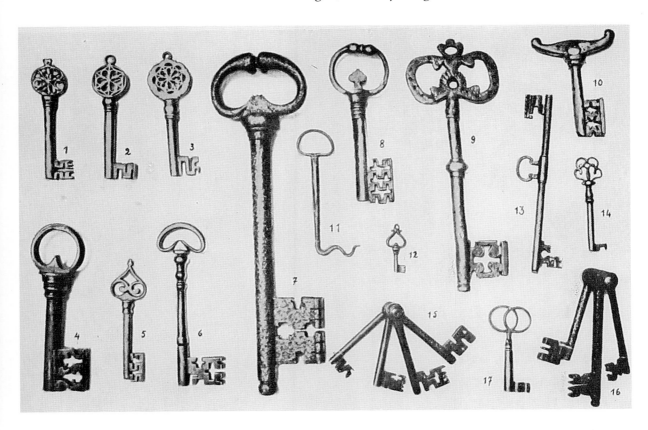

85. In Museum "Het Vleeshuis" at Antwerp.

Collection of Dr. E. Vita Israëls, Amsterdam

By the kind permission of Dr. E. Vita Israëls I reproduce in Fig. 86 his exceptional collection of keys. Dr. Vita Israëls, a chemist by profession, and a keen collector of antique furniture, clocks, paintings and china, has presented his collection of keys to the Royal Physical Society of Amsterdam.

As it would lead us too far afield to analyse the great variety of keys from the different ages, we proceed to another collection.

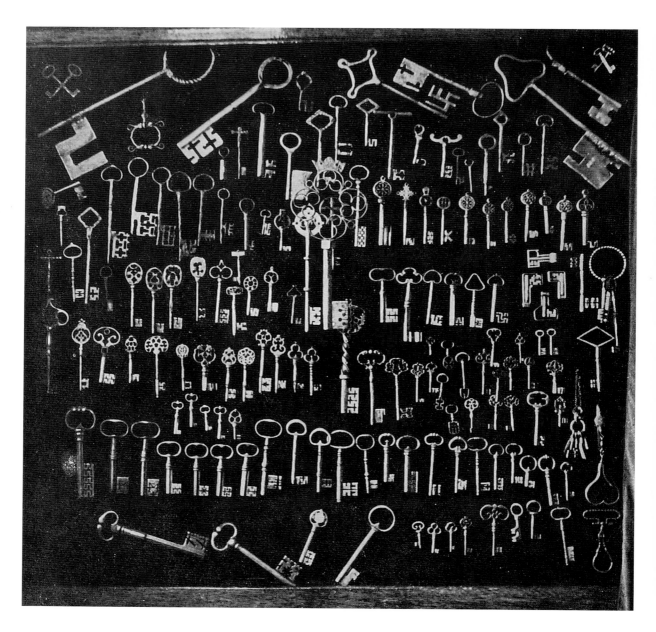

86. Collection of Dr. E. Vita Israëls, Amsterdam.

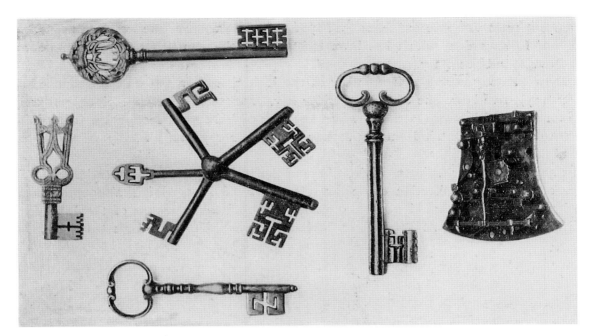

87. *Gothic key of the 15th century.*
A French Renaissance key (17th century).
Five keys, one of them a slider key, connected by a joint pivot, probably of German or Flemish origin. Chamberlain key with gilded copper monogram J. C., of the early 18th century.
Dutch Renaissance key (17th century).
Richly ornamented chest lock.
Collection Museum Princessehof Leeuwarden (Holland).

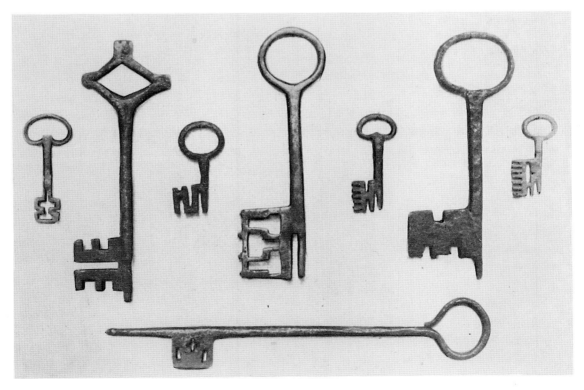

88. *Iron keys of the 14th and the 15th centuries.*

A limited but most valuable collection of keys has found shelter in the wonderful museum Princessehof Leeuwarden, which has been built up by the conservator of the museum, Mr. N. Ottema, public notary.

Fig. 87 shows, among others, a Gothic key of the 15th century, a French Renaissance key (17th century), five keys, one of them a slider key, connected by a joint pivot, probably of German or Flemish origin, a chamberlain key with gilded copper monogram J. C. of the early 18th century, a Dutch Renaissance key (17th century) and finally a richly ornamented chest lock.

Fig. 88 represents some specimens of iron keys of the 14th and the 15th centuries. Furthermore the Lips' collection comprises a number of keys, which although not a hundred years old yet, should not be omitted. The Australian agents of the Lips' factories, Messrs. Wm. Bedford Limited of Melbourne, wished to contribute their share to this collection.

89. Iron keys, Gothic and early Renaissance. (15th-17th centuries).

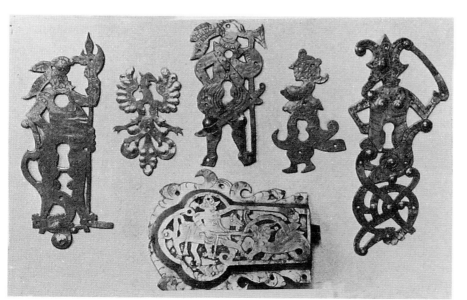

90. Swiss cupboard lock and German tin coated lock plates. (16th century).

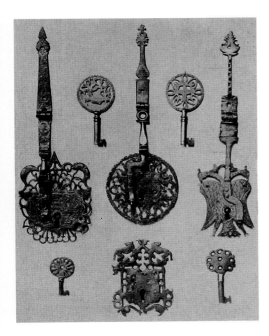 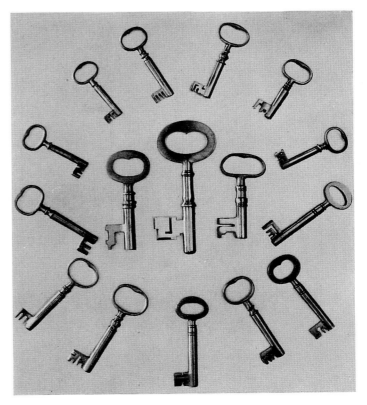

91. An outstanding collection of Spanish locks and keys of the 16th -17th centuries. Lips' collection.

92. Steel keys of English origin used for the cell doors of the Victorian Prison in Melbourne. Lips' collection.

This set of keys dates from the middle of the last century and is composed of some specimens of fine ward keys to fit keyholes of different shapes. They are of English origin and have been made of forged and hardened steel (Fig. 92).
It is recorded of the large key in the centre of the illustration that it once served as the master key of the cell door of the Victorian Prison in Melbourne and had been used as such since 1850.
Finally, collections of locks and keys are found in many other museums throughout the Netherlands, for instance, in the Kam Museum in Nijmegen, the Municipal Museum of s'Hertogenbosch, the Museum of Middelburg. Most of these keys have been dredged from the large rivers in this country.

In the history of the Roman Catholic Church the key has always played a very important part as a symbol. Christ handed St. Peter, one of His twelve Apostles, the key of the gate of Heaven. St. Peter's emblem is the twin keys symbolizing his prerogative of keeping the doors of the Kingdom of Heaven.
In many parts of the Holy Scripture the key is referred to as a symbol of legitimate admission or of admission by special consent and as a sign of dignity. St. Gervase's famous key, kept in the treasury of the Ancient St. Gervase Church, Maastricht, Holland, should be considered in this light. Dr. P. T. Kessler, Conservator of the Art

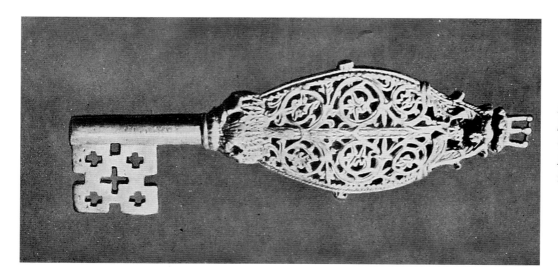

93. The key of St. Gervase of Maastricht (Holland).

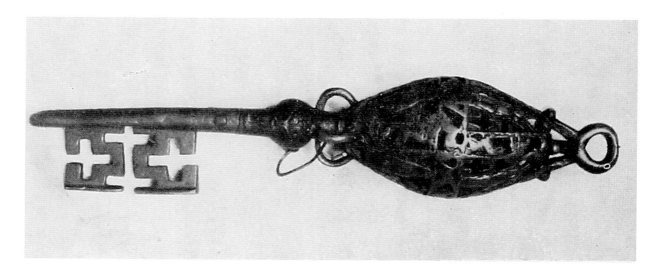

94. St. Hubert key of Liège (Belgium).

95. St. Hubert key of Liège (Belgium).

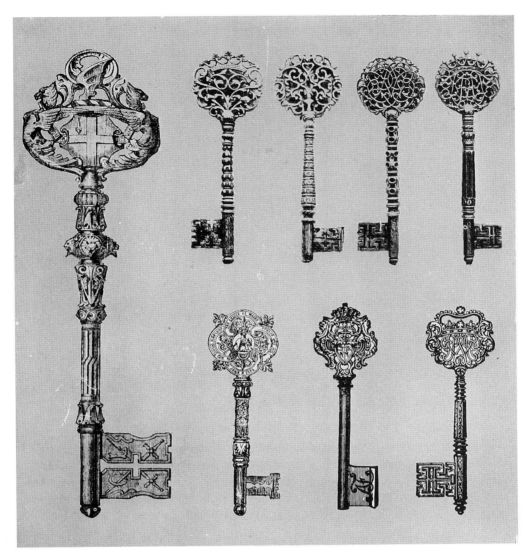

96. English ornamental keys of the 15th to the 18th centuries.

Museum in Mainz, gives the following historical details on this basket shaped key, conspicuous by its highly artistic ornamentation and design, and a masterpiece of the ancient locksmith's art.

"*St. Peter's key, made of electrum, an alloy of gold and silver, is of a highly artistic design. It has a basket shaped bow and the interstices in between the stakes have been richly interlaced with graceful slender tendrils. The 'bottom' of the basket has the form of a spherical knob which in its turn is connected to the key shank. This remarkable key, the overall length of which is about 12 inches, is said to have been found in St. Gervase's tomb in 726. Legend has it that this key was presented to Bishop Gervase by Pope Damasus in 376 A.D. as a high distinction on the occasion of a Pilgrimage to Rome (Fig. 93). Likewise St. Hubert's key, which dates back from the 7th century, was presented to St. Hubert also as a distinction, and is kept in Liege (Belgium) (Fig. 94 and 95)*".

Numerous collections in British museums, to name a few, the Victoria and Albert Museum, the British Museum, Marlborough House, the London Gallery, the Scottish

97. Musée de Cluny, Paris, German Gothic, 15th century.

Museum, the Museum of Antiquaries in Edinburgh, the Kensington Museum, the Birmingham Museum, and many more, contain exquisite keys, some of which are illustrated in Fig. 96. Among them we find the gold key with which the Prince of Wales performed the opening ceremony of the new town hall in Bolton and the key of the Temple Bar City Gate of London. Fig. 97-103 show a number of fine locks from different countries as preserved in the Musee de Cluny in Paris and in the Lips' collection. The art of chasing and embossing, as well as artistic locksmith's work had highly developed in France, England, Germany and other southern countries. Evidence can be found in practically all of these countries. The iron work of the western door of the Cathedral of Notre Dame in Paris is a real masterpiece of world fame for its symbolic ornamentation of a richness and an abundance that many a legend has originated from it. This exceptional work of art has even been ascribed to the devil himself. Striking examples of great skill in chasing, embossing and casting are shown in Italian and Spanish museums, where beautiful specimens of lock plates, locks and keys with symbolic representations cast in bronze, silver and gold are exhibited.

Some illustrations of lock plates on the following pages speak for the great progress already made in mediaeval times in the art of metal working.

On the occasion of one of his lectures on ancient and modern locks and keys in a country town in Holland, the author met Mr. J. D. Carstens a master smith of Deventer, who presented him with a section of an old tool, now added to the writer's collection. This instrument, which is adjustable in horizontal and vertical directions, was fixed to a treadle lathe and provided with a clamp to hold chisels of different sizes and profiles to cut the wards, steps and grooves into the key bit. This kind of tool, which formed a part of the kit belonging to Mr. Carsten's great grandfather, is getting rarer and rarer and is, therefore, to be considered a valuable asset to the Lips' collection (Fig. 106).

98 - 99. Musée de Cluny, Paris, 15th century.

100 - 101. Musée de Cluny, Paris, 16th century.

102-103. Musée de Cluny, Paris, 15th century.

104 - 105. Museo Nazionale (il Bargello), Florence Italy.

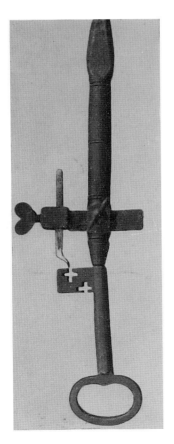

106. Section of an old tool from Mr. J. D. Carsten. Lips' collection.

107. Museo Artistico Industriale, Rome.

Ancient Locks Collected in Museums

The reader will have understood that respectable collections of locks and keys have been concentrated in museums in many parts of the world, not to speak of private collections. It is impossible within the scope of this book to detail all of them, nevertheless the following pages contain a number of good reproductions, particularly of ancient locks of French, German, Italian, British and Dutch origin, and many of them are represented in the Lips' collection.

From a constructional angle these locks show considerable differences. Some are doublethrow locks with a single bolt, which is deadlocked by an extra turn of the key, some are double bolt locks with a spring and a dead bolt or a day and night bolt. Even three- and four-bolt locks are among them, with the usual spring and dead bolts and additionally a bolt operated by a knob or handle.

All these types of locks were principally applied either to main entrance doors of buildings, or to chests and cabinets and in some cases to steel strong boxes.

It is safe to assume that in the olden days German art smiths already rivalled their French, British and Flemish colleagues with their products of superb design and of really artistic value, particularly in the field of chase work for gates and ornamentation of doors.

Clock and watchmaking is characterized by its high precision and the products of the clockmaking industry of Thuringia and the Black Mountains in Germany are acknowledged by the whole world as examples of real craftsmanship. Now the fact that many lockmakers have sprung from the clockmaking branch signifies that these crafts have much in common, and both crafts afford an equal opportunity to display the skill and sense of beauty of those craftsmen.

Jörg Heusz ranked first among the best known clock and lockmakers and received the title of Master Clock and Lockmaker in 1449, at the same time the freedom of the town of Nürnberg was offered to him. One of Jörg Heusz' outstanding creations was the exquisite clock of St. Mary's in Nürnberg on which the Seven Electors appear in succession. His lockmaking achievements were nonetheless famous and specimens of great beauty were made by him for many municipal buildings as well as for private dwellings. Hans Ehemann, also of Nürnberg, deserves mention here as one of the leading craftsmen, not only on account of his exquisite ornamentation, but also because of the highly ingenious design and construction of his locks. He is said to have been the inventor of the letter ring padlock and Daniel Schwerter (Nürnberg 1651) records that the use of this type of lock was not restricted to Germany alone but, on the contrary, these locks were exported to other European countries. Such was his reputation that he was invited to make special locks for the French and Italian Courts.

Finally, Bartholomew Hoppert (1648—1715) of Roth must be considered as an artistic lockmaker of excellent reputation. Hoppert spent many years in Holland, as well as in other European countries, where he produced locks of magnificent design. He was invited by King Louis XIV of France, a great patron of art smith's work, to make special locks, which task Hoppert carried out to his great credit. He returned to Germany in 1677 and many of his products can be found in the Art Museum of Dresden.

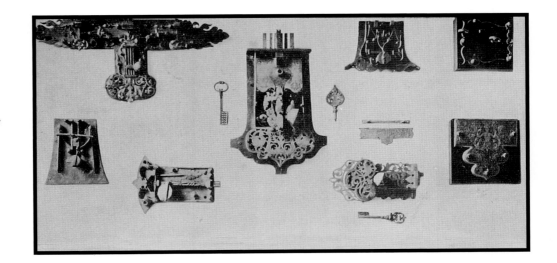

108. Germany 16th-17th centuries.

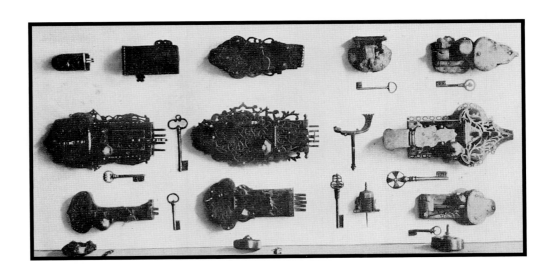

109. Germany 16th-17th centuries.

110. Germany 16th-17th centuries (A selection from collections English Museums).

111. Germany 15th-16th centuries.

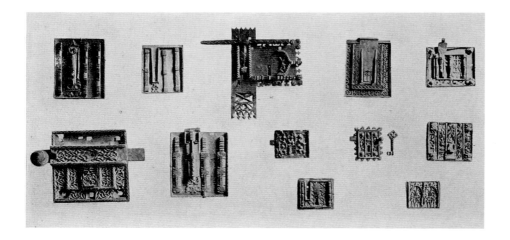

112. France 15th century (A selection from collections in English Museums).

113. France 15th century.

114. France 16th-18th centuries.

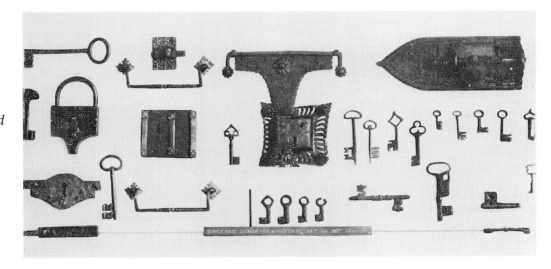

115. England 16th-18th centuries.

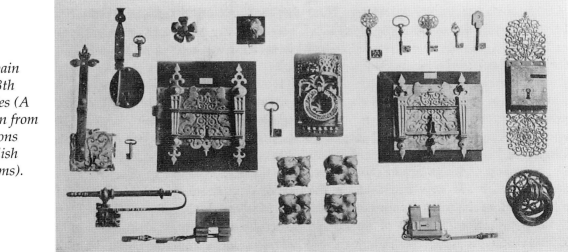

116. Spain 16th-18th centuries (A selection from collections in English Museums).

Fine collections of locks are also displayed in the National Museum in München, Germanisches Museum of Nürnberg and the Städtisches Museum in Nürnberg. In all probability these collections are not of particular interest to the average sight-seer but to those conversant with these crafts they convey a lot, as locks and keys have been in use from times immemorial and can therefore be said to represent the oldest appliances of mankind. An exceptionally fine collection is Andreas Dillinger's in Vienna to be found in the Technologisch Gewerbe Museum in Vienna. Andreas Dillinger was a well-known collector and connoisseur of old wrought iron work, and his collection was protected by the Government. Since the war it has partly been sold to different countries.

It can be seen that practically every country has its own specialists in lock making, all enjoying international reputations.

The Gothic style of wrought iron work was mainly applied in the 13th and 14th centuries, and to a lesser degree in the 15th century, and its ornamentation mostly consisted of symbols taken from the vegetable and animal kingdom. In the early Gothic style these symbols still show rather primitive features, however, when this form of symbolization had reached perfection more natural features developed, and in the late Gothic period even became over elaborated. These symbols are also found in iron work on doors, locks and furniture and principally consisted of lilies, roses and thistles.

Likewise Renaissance ornamentation depicted symbols from the vegetable world, especially leaf-symbols of exotic flowers, of which the tendrils gradually branch off into spirals and scrolls, whilst symbols from the human and animal world are frequently found. The magnificent town hall of Mons exemplifies a masterpiece of 15th century architecture. It has a double folding main entrance door fitted on the inside with a massive brass lock of a solid construction. An escutcheon of artistic pattern covers the outside keyhole, as shown in Fig. 158. It was designed by an apprentice of the smith's guild as a trial piece to gain his title of master craftsman and depicts the coat of arms of the town of Mons, viz. two towers and a draw bridge. To symbolize safety and security a watchdog was placed just below the keyhole. Note the heavy door knocker, mounted close to the escutcheon. An Angel appears on the top of the knocker, thus symbolizing vigilance. This lock and knocker, made completely of brass, are no longer used and serve as decoration only.

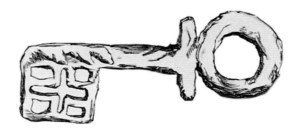

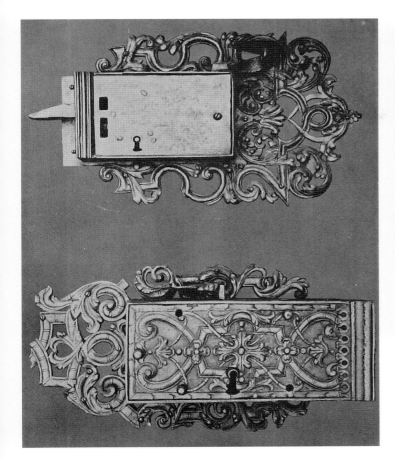

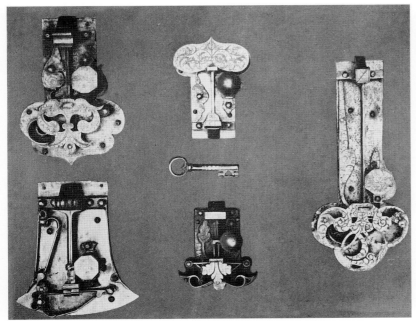

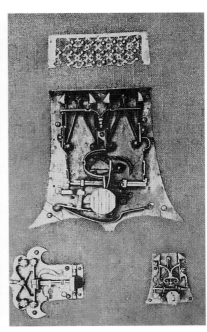

117. 17th century Front door locks - Lips' collection.

118. Front door locks. A selection from German Collections 16th century.

119. Chest and trunk locks of the 16th and the 17th centuries. A selection from collections in German Museums.

120. Chest locks.

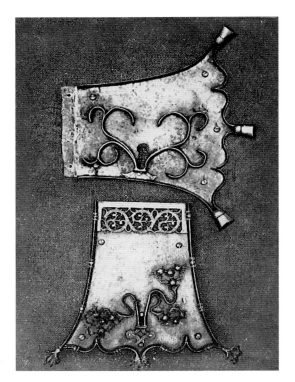

121. Chest locks.

122. Chest locks from different German Museums 16th century.

123. 16th century South German or Tyrolese Trunk lock.

124. Early 16th century German Chest lock.

125. Italian Chest lock 15th - 16th centuries.

126. German Chest lock 15th - 16th centuries.

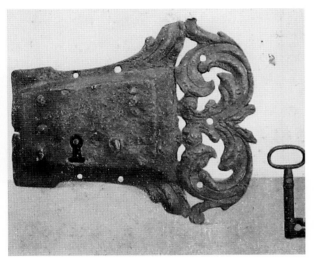
127. French door lock – Louis XIV. The key to this lock was made at a much later date.

128. 17th century German or French door lock.

129. German Front door lock 17th-18th centuries.

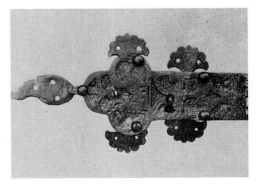
130. Flemish front door lock 16th century.

All locks and keys shown on this page are in the Lips' collection.

131. *German or Flemish front door lock (17th century).*

132. *French front door lock (early 18th century).*

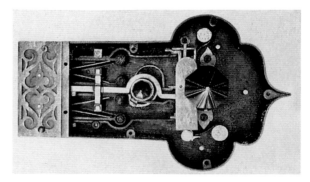

133. *German front door lock 17th century.*

134. *Flemish front door lock 17th century.*

135. *French front door lock 17th century.*

All locks and keys shown on this page are in the Lips' collection.

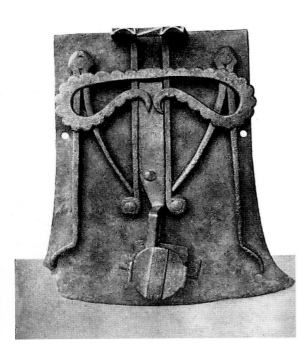

136. German chest lock (14th-15th centuries).

137. Flemish chest lock (14th-15th centuries).

138. Flemish trunk lock (latter part 16th century).

139. English chest lock (middle 16th century).

140. French chest lock (16th century).

141. Shackled bolt lock (14th-15th centuries).

142. French or Italian chest lock (15th-16th centuries). Both specimens are in the Lips' collection.

143. Door lock (16th-17th centuries).

144. Dutch door lock (17th-18th centuries).

145. Flemish chest lock (16th century).

146. Dutch door lock (17th century).

147. Dutch door lock (17th century).

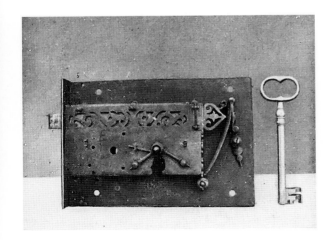

148. Dutch door lock (17th century).

149. Flemish door lock (17th century).

150. Lock from Sieburg near Cologne. Ornamentation in gilded brass - (about A.D. 1750).

151. Church door lock of St. George's at Dinkelsbühl (Bavaria), (about A.D. 1750).

152. German chest lock. Engraved and painted, (about A.D. 1700).

153. Hinged lid of ancient strong box (about A.D. 1600) In the Städtisches Museum, Nürnberg.

154. German chest lock of the 15th century. Richly ornamented with tendrils in brass. Bayerisches National Museum, Munich.

155. German door lock (17th-18th centuries).

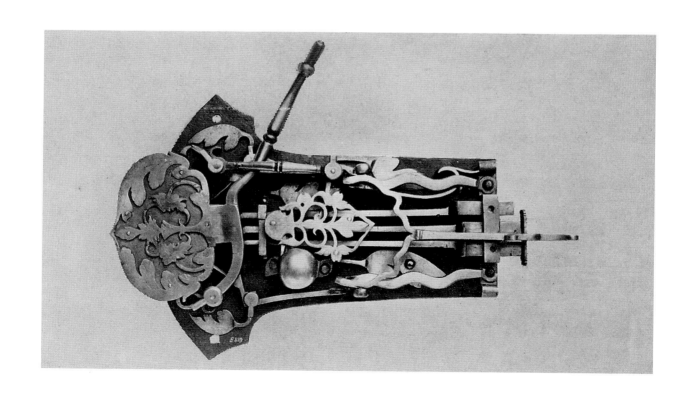

156. German door lock (17th century).

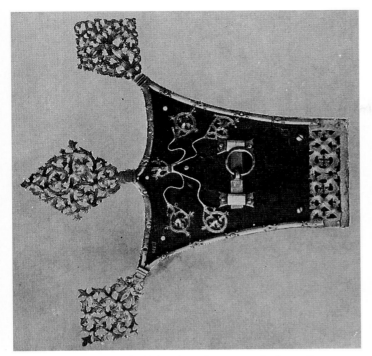
157. German chest lock (end 15th century), from the Bayerisches National Museum, Munich.

158. The lock of the main entrance of Mom Town Hall (Belgium).

Collection in the National Museum, Amsterdam

A limited but valuable collection of ancient locks and keys, a number of which are reproduced in Fig. 160, can be seen in the Rijks Museum in Amsterdam.
The lock shown in Fig. 159 is by far the most outstanding specimen in this collection, since it was the front door lock of the Mauritshuis in The Hague, in the 17th century. Similar locks were used also as front door locks for other manors and castles in Holland. The miniature man at the back of the brass lock case covers the keyhole with one of his legs. The leg is extended by pushing a secret button or catch, and thus the keyhole is released to receive the key. However, the moment the leg is stretched, the toe-cap of the little fellow's boot points to a certain number on the dial. By this dial the number of times the door has been opened can be ascertained exactly, as each turn of the key causes the number on the dial to change. The lock bolt can only be released by pushing back the little fellow's hat and when this has been done the key can perform its normal function of opening. A remarkable lock indeed, made by John Wilkins of Birmingham. The same collection also comprises a gilt-brass padlock of about the year 1700, a number of iron keys of the 15th, 16th, 17th and 18th centuries, two ornamental keys, one of which bears the coat of arms of the Elector Franz Georg von Schonborn, Archbishop of Trier. There is also a gilt-brass chamberlain's key showing the coat of arms of the Prince-Bishop of Wurzburg and the Bishop of Bamberg, F. L. von Erbach; the forged iron key of Clemens August, Elector of Cologne (1725-1764) is a specimen of remarkable beauty. A reproduction of this remarkable key is found in Fig. 161.
Finally, the collection contains an English ornamented steel key, bearing the initial C.M.

159. Front door lock in the 17th century of the "Mauritshuis", The Hague and other castles in Holland. Lips' collection.

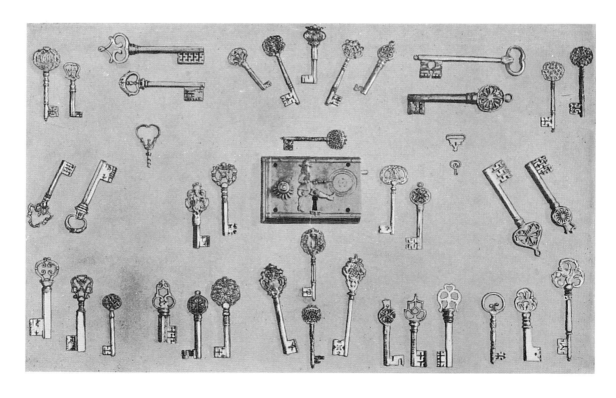

160. Collections of English ornamented steel keys. (Rijks Museum Amsterdam).

(1700) and an escutcheon showing the coat of arms of Gorkum or Buuren and the date 1587. The valuable escutcheon and lock reproduced in Fig. 162 were presented to the writer from a private collection. The wrought iron escutcheon, shown at the top of this illustration, is of Italian-Gothic design and served to cover the keyhole of a secret wall safe lock. By pushing a button, in the centre, the nicely carved and chiselled horse and its rider turn away, freeing the keyhole.

The lock in the picture of Fig. 163 is of Swiss origin in Louis XIV style. At the back of this lock, which has been made of brass throughout, the following inscription is engraved : *"Virtus Nobilitai, Juste, Sobrie et Religiose Posteritati Recomendate"* (Virtue ennobles; Justice, Sobriety and Religion be recommended to posterity). Indeed no better place could have been chosen for this maxim, which could not be overlooked by those opening the lock daily.

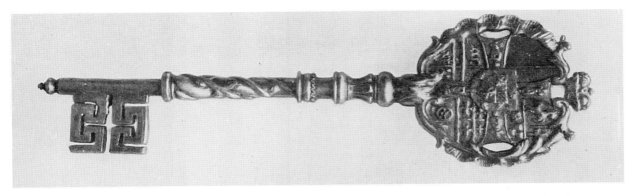

161. Ornamented key with bow showing the coat of arms of Clemens August, Elector of Cologne.

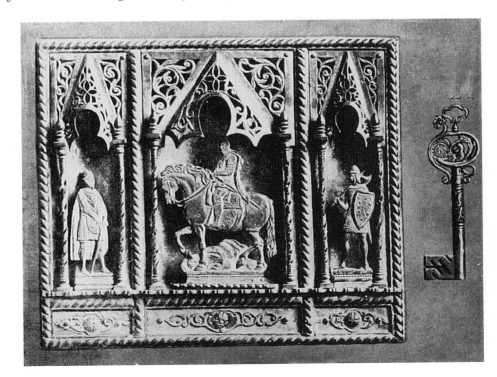

162. Valuable escutcheon of Italian-Gothic design. Lips' collection.

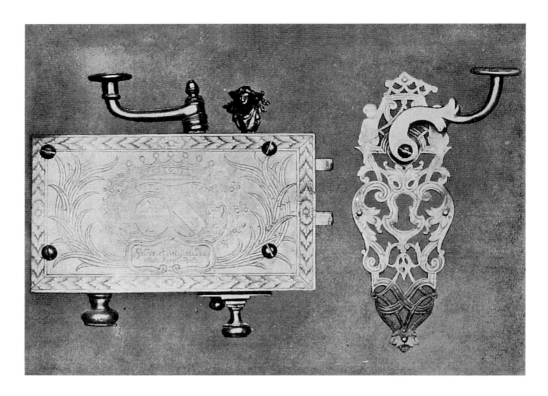

163. Lock of Swiss origin in Louis XIV style. Lips' collection.

Ancient Padlocks

Since padlocks of diverse shapes and constructions have been used for many centuries, further space will be devoted to their historical background and development throughout the ages. In the old days they were mainly used to lock iron and wooden chests, to secure town gates, prison doors and the like, and also to lock the shackles of slaves.

In the preceding chapters we have been confronted with scores of locks, vastly different in model and construction. Considering the primitive tools available in those days, we often wonder, and rightly, how our forebears managed to produce such masterpieces of ingenuity. The question arises: In what manner was security obtained? An answer can only be given in general terms, as a detailed survey of ways and means would mean exhaustive information on designs, construction etc. which it is not intended to supply in this book. Fig. 164 shows some ancient padlocks from the Lips' collection.

In olden times ornamentation played a very important part in the construction of locks and keys and consequently excessive attention was paid to the outward appearance of both locks and keys. It would appear that the success of a lock design was determined mainly by its ornamentation; indeed little progress was made in the construction of real security locks between the Roman period and the middle of the 17th century (Figs. 165 and 166). Many locks were provided with obstacles in or around their keyholes in order to obtain a maximum security, which safety measure was generally applied in all countries. All locks were made entirely by hand and so had to be considered bespoke work.

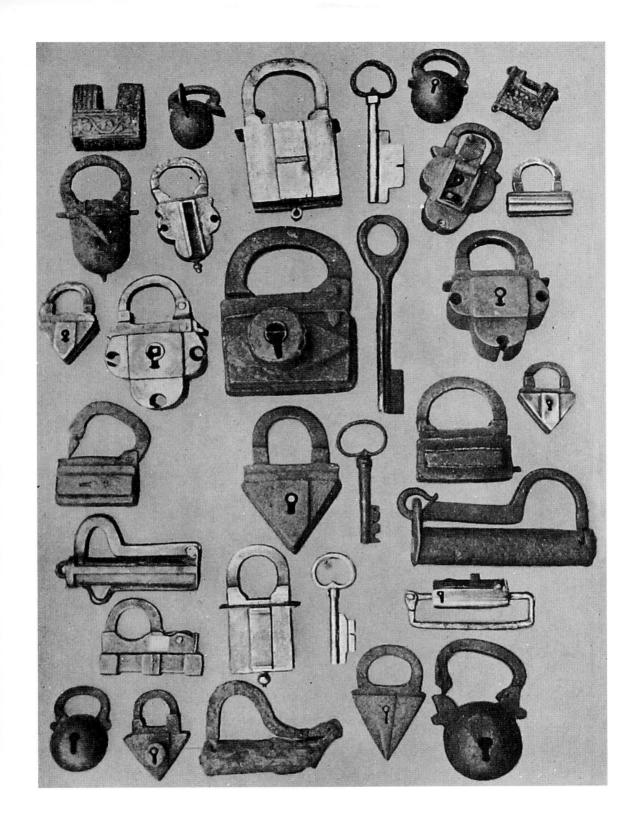

164. Ancient Padlocks in the Lips' collection.

165. Old-Spanish padlock for town gates. Lips' collection.

166. Old-Spanish padlock for town gates. Lips' collection.

167-168 Ancient Money Safe from China. Opened and locked. Lips' collection.

The design and construction of every new lock portrayed the momentary caprice of the lock maker, viz. a complete indifference to standardization, or uniformity in dimensions, design and construction. In fact no real desire to achieve maximum security. The blacksmith, who quite rightly made the construction of gates, arms and signs his speciality, also acted as "lockmaker" to the detriment of security, and left much to be desired from that angle. Gradually the skill in working metals, iron in particular, progressed and extended. This led to the construction of iron chests and safes to protect documents and books and to keep them under "lock and key", which widened the field of application for locking devices beyond the use of locks and keys in buildings.

Ancient Chests and Boxes

The Lips' collection contains some valuable specimens of ancient chests and boxes, pictures of which are shown in Fig. 167 and 168.
The reproductions at the top represent an ancient safe in locked and unlocked position, which was used for many years in a Chinese Bank in the Dutch East Indies. The lock mechanism is operated by three different keys; the largest key throws the bolts, whilst the remaining two keys serve to block the main lock. The illustrations in Figs. 169 and 170 show a beautifully decorated old Dutch strong box, the lock of which is operated by a heavy key.

On the inside of the cover there is a sheet iron plate with a fine open work pattern, which exposes the lock mechanism. The chest is complete with the two padlocks, which are very rare.

The steel doors shown in Figs. 171, 172 and 173 are fitted with locks of Dutch origin showing remarkable constructional features.

Fig. 171: This lock mechanism, which is a combined key operated and letter combination lock, is covered by an open work brass plate, which exposes the mechanism.
The bolt mechanism is released by a Bramah key only after the slots of the four knob letter lock have been placed in alignment and the heavy middle knob pushed downwards. This combined lock, made by J. C. de Graaf, an Amsterdam Master Locksmith, was made entirely by hand, and dates from about 1830.

Fig. 172: This door has two secret flaps which open by pushing a secret button, after which the key can be inserted. This door was made by an Amsterdam Master Locksmith, B. Rensing, in 1814.

Fig. 173 represents the lock and key of a tabernacle door in a Roman Catholic Church. The key wards bear the initials JHS which are repeated in the lock to protect the bolt mechanism. This lock dates from the 19th century.

169-170. 16th century strong box with padlocks. Lips' collection.

171. Door with letter combination lock (A.D. 1830). Lips' collection.

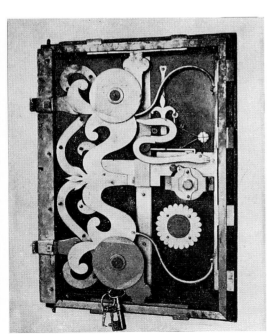

172. Steel door fitted with locks of Dutch origin (1814). Lips' collection.

173. Lock and key of a tabernacle door in a Roman Catholic Church (19th century). Lips' collection.

CHAPTER IV
Locks and Keys and their Manufacture

The design of the key predetermines the design of the lock mechanism, which means that the key is made first and afterwards the lock. Subsequently the key bit constitutes the principal part of the key, since this part brings about the locking and unlocking of the mechanism. A complicated key bit besides precision as regards the axial movement of the key on its pivot is a determinant of the degree of security in the lock.

Let us first discuss the most popular type of lock, viz. operated by a turning key.

Locks may have one, two or even more bolts, which can be thrown or withdrawn by turning the key. In the case of locks of an older construction two or more turns of the key were required to shoot or withdraw the bolt over its full length.

In South European countries locks are still being generally used of which the bolt is fully thrown or withdrawn only after six or even eight turns of the key. The reason for this may be found in the lack of wood of a suitable quality to make door posts sufficiently resistant. In Italy locks are still found of which the bolts are fully thrown or withdrawn only after three to eight turns of the key. As we shall see later, multiple throw lock bolts are useless in modern locks, in which either one long throw or a double short throw bolt is applied. It should be noted that with every extra turn of the key the same action is consecutively repeated in the lock mechanism.

To illustrate the operating principles of a bolt mechanism, a sketch is given in Fig. 174 of a very simple double action lock.

The bolt of a lock is composed of two parts, viz. the bolt head and the bolt tail. As a rule the bolt head is thicker and of a more solid construction than the bolt tail and is guided by the hole in the front plate of the lock. The bolt tail, on the other hand, is supported by studs riveted to the lock case. After the key bit has entered the lock through the keyhole, the nose of the key bit immediately engages a talon—a notch or gap in the bottom part of the bolt tail; by turning the key the bolt will slide and is either thrown or withdrawn. The position of the bolt, whether thrown or withdrawn, should always be such that the key bit catches the talons in the right place. To accomplish this the upper surface of the bolt tail is likewise provided with notches which coincide with a spring loaded pin tumbler or fence. This spring loaded tumbler prevents the movement of the bolt until the tumbler is lifted by an additional turn of the key, either to throw or to withdraw the bolt. After this operation the pin tumbler drops again into one of the notches. This simple action explains the functions of the parts of a lock mechanism providing the security elements of a key operated lock.

Moving the bolt by other means than the proper key was prevented by providing all kinds of obstructions in or around the keyhole, so that only the key bit corresponding to the talons and notches in the lock mechanism could enter the lock and steer clear of the obstacles when turning the key. However, the security obtained by wards soon turned out to be inadequate, so that lockmakers proceeded to bush the centres of the key holes with steel bushes of different shapes, which allowed only the correspondingly shaped key to enter the lock. Some of these bushes are illustrated in Fig. 175.

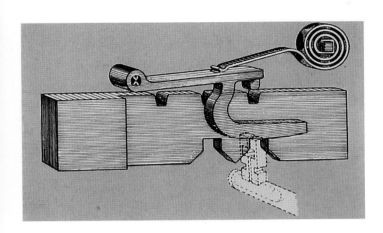
174. Operating principles of a bolt mechanism.

175. Forms of key holes in old locks.

176. Italian chest lock (16th century). Lips' collection.

177. An ancient French warded chest lock and key. Lips' collection.

178. Specimens of keys in the Lips' collection.

179. Specimens of keys in the Lips' collection.

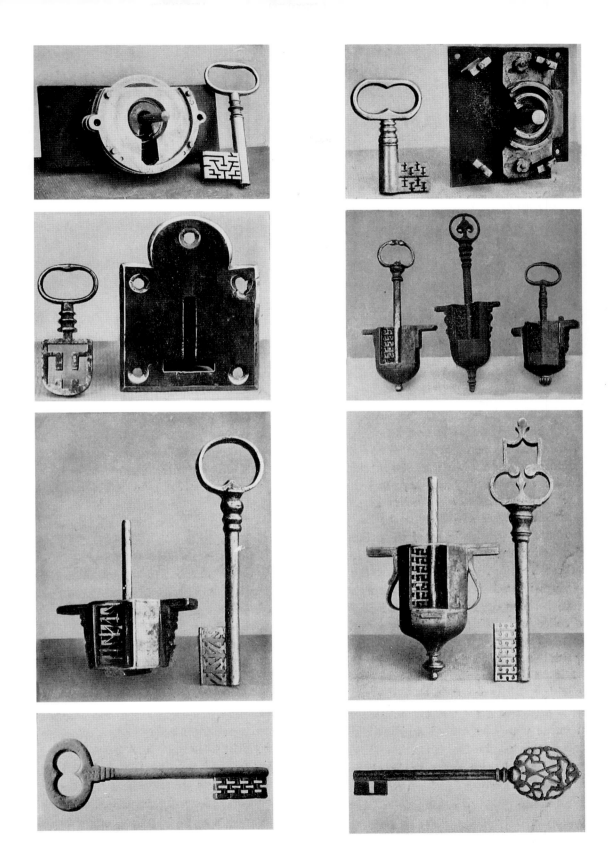

180-187. Warded and bushed locks and their corresponding keys in the Lips' collection.

They clearly indicate the endeavours of lock manufacturers to resist burglary.
Indeed these bushings greatly enhanced the security of locks, as in locking and unlocking such bushings moved synchronously with the turning key, protected the wards and improved the bearing of the key. Some fine specimens of bushes of a complete design in the Lips' collection testify to great craftsmanship and love of the craft. The reproduction of a 16th century Italian chest lock (Fig. 176) clearly shows the security device, a keyhole with the correspondingly shaped key pipe to match. This ancient lock of the Lips' collection is the finest specimen the writer could lay his hands on. The key nose shows further a profile with incisions or wards, which again have to correspond with the incisions of a stud in the lock to turn the key. The construction of the key pipe and the corresponding bush and stud is a feat of great mechanical skill and precision. Furthermore this reproduction gives an inside view of the lock and reveals the security mechanism, the studs of which have been riveted to the lock case. The key of this lock was made of hardened steel throughout to avoid wear and tear. In all probability this lock served some extraordinary purpose, for a royal crown has been engraved in the lock plate. It may have been made by a lock smith to qualify for his title of Master locksmith. The writer acquired the lock at an auction in Italy.

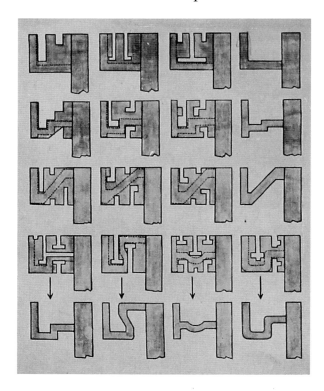

188. Warded keys and false key.

In later years bushed keyholes were also used regularly and even lever type locks of modern construction are based on this principle. Several examples are shown in Fig. 179. The shapes of the key bush and wards resemble the similarly shaped wards and bushings in and around the keyhole. This part of the key collection also comprises specimens of the Bramah type, one of which is shown in the left bottom corner of the reproduction, specimens of the Bramah-Vago type of Italian origin, of the Bramah-Chubb type from England and some of the original Bramah types. These keys are flanked by some fine handmade Church door keys. The illustration in Fig. 177 represents a French warded lock of particular design, whilst several warded and bushed locks and keys for various purposes have been reproduced in Figs. 180 to 187.

Shapes of key bushes generally allowed of very few variations, and therefore a wider scope of such indispensable variations was mainly found in the design of the wards or incisions in the key beard.

For years and years warded locks were widely used and held their own in ordinary internal and external doors until about 75 years ago. Eventually they fell into disuse for the simple reason that this type of lock no longer offered the required standard of security. Strong competition among manufacturers necessarily led to price cutting and this inevitably affected the quality of their products, which no longer deserved the name of precision instruments. The modern lock industry has however gradually moved in another direction, the cheaper grade warded lock is still the mass product of the German, Dutch, British and American lock industries, although to-day the security elements in locks of an average quality are provided by the lever mechanism. The complete inefficiency of warded locks in our modern times has been finally proved, since this type of lock can be easily operated by a key of a very simple design and construction, which keeps clear of the wards. In Fig. 188 a series of warded keys are shown, each row of keys is flanked at the right by the corresponding lock pick of which the bit has been cut in angles covering the basic angles of the genuine key. No matter how complicated the shapes of the wards of key bits may be, such lock picks easily open any of the corresponding locks.

It is to be regretted that a great many people place their entire confidence in this cheaper class of key, which although of a complex design, does not offer any reasonable security. Sheer ignorance on the part of the public accounts for this, the majority of people have never even glimpsed the lock mechanism and its operating parts, neither have they really studied the functioning of those parts. Moreover, qualified and expert knowledge of lock security belongs to a small minority. Nowadays many locks have keys displaying intricate wards to make them a better selling proposition, but in reality they are far from secure. It is, therefore, the duty of the Law to protect the Public against such practices, as was done in the days when this craft was protected by the Government.

The chapter *"Punitions des Cambrioleurs dans l'ancien temps"* from the 1911 publication entitled: *"Manuel de Police Scientifique"* by Professor R. A. Reiss of Lausanne, reads: *"Le vol au moyen de fausses clefs était puni très sévèrement dans l'ancien temps. Il en était de même pour la vente de fausses clefs.. Encore le 7 juillet 1722 Jean Lamy, pour avoir vendu des fausses clefs, fut rompu vif. Le 21 du même mois, Jacques Belleville, pour le même fait fut pendu"*. Similar punishments are recorded in Bally's work: *"Cambrioleurs et Cambriolage"* (Paris). Every human being has an innate tendency to safeguard his property; the reader will recall that even when a child he had this tendency to protect his toys from his playmates. Naturally man's desire to safeguard what he owns is intimately bound up with the most efficient means of achieving this protection.

Tradition has it that seals were applied in many cases in the very early days of our era. Darius commanded Daniel's lions den to be sealed, so that the King's seal made a lock superfluous. Even to-day the value of seals is recognized and respected; Governments make ample use of seals for important documents and so do the Post Office and other Public Institutions.

However, seals as well as inferior types of locks are of no use for doors. Anyone visualizing the high degree of security obtainable with precision locks, products of this modern mechanized age turned out by reliable companies fully conscious of their responsibility to the community, will be unpleasantly struck when having to face collections of low grade locks as well, thus confusing the average buyer, who through lack of experience cannot separate the chaff from the wheat. It should be remarked here that expensive first class security locks cannot be used for every conceivable purpose and it can happen that a buyer purchases a lock, or something resembling one, only to find out later that he has landed himself with a blind bargain. Negligence also accounts for this exploitation of the unwary buyer. For instance when purchasing a knife, a watch or a car the prospective buyer will scrupulously test and scrutinize the object before finally purchasing it, but whoever heard of or witnessed such earnestness in acquiring a lock.

So long as the lock fitted to the door has not been forced by burglars the owner is quite satisfied. Undoubtedly a more enlightened knowledge of the skill and methods of "the gentry" would increase the demand for genuine security locks. But not all the blame must be laid on the users. Competition in the lockmaking industry inevitably means a sacrifice to quality. In many cases the outward appearance is all that matters, but there are really good locks available for the more discriminating and intelligent buyer.

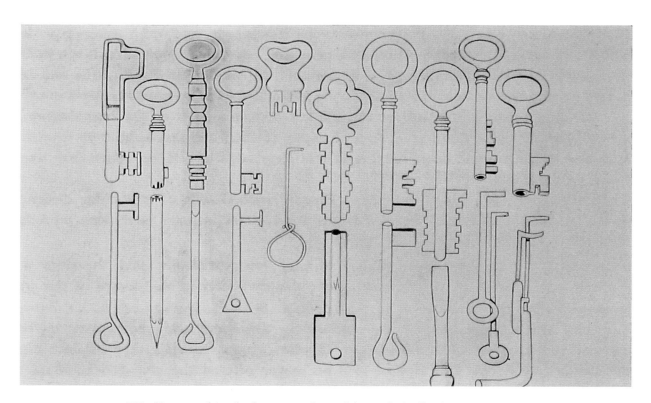

189. Keys, and in the lower section, picks and similar instruments.

CHAPTER V
Modern Locks and their Applications

Lever Locks

By the middle of the 18th century the British Isles could boast of an important and prosperous lock making industry in and near Wolverhampton, where every effort was made to attain a maximum of security. Since then considerable technical progress has been made and locks and keys have substantially improved: thousands of patents have been secured, of which many have proved valueless and have been cancelled. Full justice is due to five ingenious inventors — Barron, Bramah, Hobbs, Chubb and Cotterill, who have all endeavoured to bring safety lock mechanisms to constructional perfection.

Barron

In 1778 Mr. Robert Barron applied for an English patent for a lock of new construction, which was granted to him. This construction was based on quite new principles in as far as the shooting mechanism of the lock bolt was concerned. An illustration of the lock mechanism from Price's book shows wards around the key hole, but the real characteristics of this mechanism were the two spring tumblers checking the bolt, the bolt tail being provided with notches corresponding to these spring tumblers. This invention meant a revolution in lock construction and wards around the keyhole fell into disuse. The spring tumblers were lifted to the proper level by the various cuts in the key bit in order to release and move the bolt. Over or under lifting of the tumblers was avoided by a proper division of the cuts in the key bit, so that the division line of the tumblers coincided when the tumblers passed from one notch to another. An illustration of Barron's twin tumbler lock shows both the slots in the bolt tail and the differently sized tumblers. However, Barron stuck to the principle of a properly designed fixed warded bush around the keyhole to protect the tumbler mechanism and to prevent the key from entering unless properly shaped. See Figs. 190 and 191.

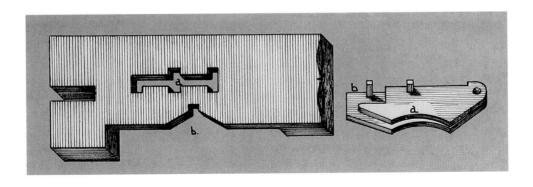

190. Mr. Barron's twin tumbler lock.

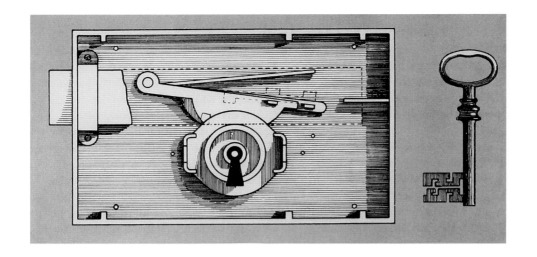

191. Mr. Robert Barron's lock (1778).

Joseph Bramah

Hardly any other lockmaker in England took out so many patents as Joseph Bramah for his revolutionary lock inventions, one known as the Bramah lock is shown in cross section in Fig. 192. Joseph Bramah organized a well planned publicity campaign for his new invention, which soon won a world-wide reputation.

Although Bramah's original invention was superseded by improved constructions, it contributed an important share to the progress in lock making. When in 1784 Bramah took out his first patent, London was alarmed by many sensational burglaries, which did not fail to impress the public, terrifying the propertied classes and in particular bankers and, of course, provided a great stimulus in popularizing Bramah's lock, the principles of which differed widely from the orthodox locking systems of his day.

The pipe of the key fitting on a drill pin is provided with differently sized grooves, shaped to coincide with the corresponding spring loaded sliders in the lock mechanism. These sliders, of which one is shown in Fig. 193, have in their turn notches in different positions. One part of the slider is recessed in the mechanism and the other part rests in a fixed ring with indentations, thus obstructing the movable parts of the mechanism. By inserting the key, the notches in the sliders are adjusted to coincide with the projections of the ring, by which action the mechanism is released and the key can be turned.

In its day Bramah's lock was known for its great security, but progress has since reduced its efficiency and it is no longer resistant to lock picking.

According to Price, Joseph Bramah was the first in the lockmaking field to adapt the manufacturing process to the technical requirements by designing and producing his own machinery. Bramah took every precaution not to reveal his manufacturing secrets. This machinery is to-day still in their workshop in London.

The lockmaking industry took over the basic ideas of Bramah's invention, and various modifications and improvements in the construction have made this type of lock extremely popular and even nowadays it is widely used in many countries.

A fine specimen of a combined high security Bramah-Vago lock is in the Lips' collection and is illustrated in Figs. 195 and 196.

Different variations in design and construction followed in the way of the Bramah-Chubb invention, of which the locking device of Robert Schneider of Dresden is outstanding. The recessed grooves in the top end of the key pipe are set by rotating combination rings. Any one familiar with double combination rings will be able to operate the lock. These illustrations show some marvellous achievements in the lock field by Vago of Milan, who succeeded in adding two further safety features. Some of Vago's locks worked on as many as 24 different combinations.

To-day the Vago Milano patent is held by Lips (Lips-Vago). Illustrations Figs. 195 and 196 show the Bramah lock of a later construction, provided with combined levers which had to be set by a key. Locks of the Bramah type were operated by a pipe key to fit on a drill pin, which engaged the matching grooves in the bolt tail.

The writer joins Price in giving credit where credit is due, for although Bramah locks have undergone severe tests, his invention has held its own.

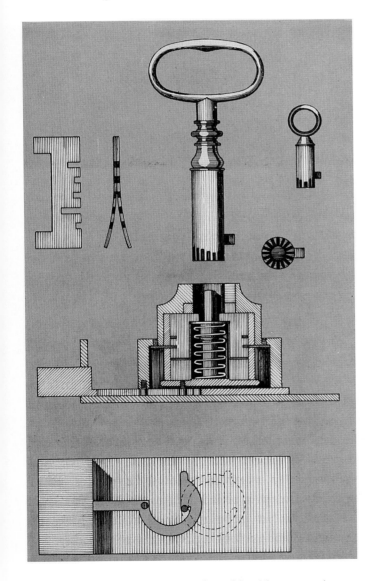

192. Bramah's invention of world-wide reputation.

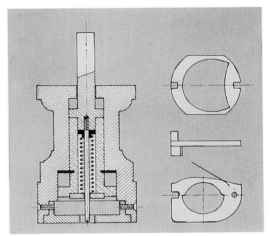

193. A Bramah Cylinder Lock.

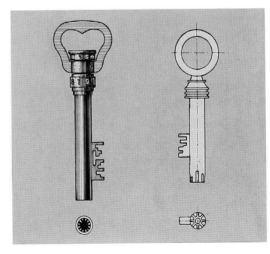

194. Schneider and Brahman key.

111

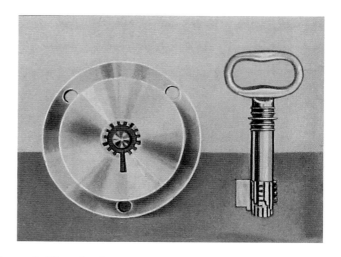

195-196. A Bramah-Vago Lock.

Cotterill

Cotterill's invention was patented in 1846, and was marketed under the title of "Patent Climax-Detector lock". The mechanism was built on the principles of the Bramah lock, the only difference being that its spring loaded sliders were positioned at right angles to the grooves in the key. Some sketches in Fig. 197 may serve to illustrate this construction. A number of spring loaded sliders in the cylindrical lock case engage corresponding radially arranged grooves; the lock case further contains a circular shaped groove, whilst the sliders are grooved as well. By inserting the proper key, provided with matching grooves, the sliders in the lock case will give way and this action will cause the groove in the lock case and the grooves in the sliders to coincide. The lock plate has a number of notches which match the grooves perfectly. Once the sliders have thus been adjusted, the notches are released and the key can be turned to move the bolt. Cotterill's lock was once picked by a Mr. Hobbs of New York, and this experience led him to add a detector to the circular shaped grooved ring by which the security of this lock against picking was considerably improved. The centre sketch of the illustration shows this detector, while the lower sketch demonstrates the key inserted in the lock.

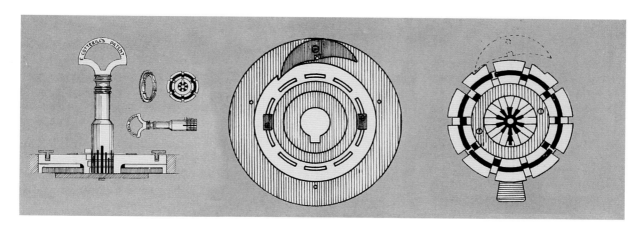

197. Cotterill's invention: Patent Climax-Detector lock (1846). Lips' collection.

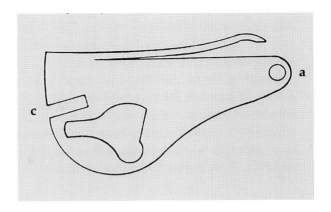

198. English form of lever.

Chubb - London

The Marquis of Worcester (1640) once saw in a dream a lock containing a device which, like a fox trap, suddenly caught the hand of any intruder trying to force it. In this way evidence of the intruder's presence was given instantly. About 200 years later the legend of this vision reached a lock maker, Mr. Jeremiah Chubb, who enjoyed an excellent reputation. In 1817 Portsmouth was shocked by a daring burglary committed with the aid of lock picks. The Government offered a reward of £100 to anyone who could produce a lock that could not be operated other than by the corresponding key. Numerous inventors presented themselves, but none of them could outdo Chubb, whose invention was patented in 1818. In the field of lockmaking the Chubbs occupied a predominant place and the book entitled *"Lock and Lockmaking"* by Mr. F. J. Butter, who for years held a leading position in the Chubb company, is recommended to any reader seriously interested in the subject. Jeremiah, Charles and John Chubb each contributed his share to perfecting their lock construction, patents for which were granted in 1824, 1833, 1846 and 1847. Their original four lever lock was provided with a warded keyhole and the faces of the levers had sham notches. Later on the detector principle was applied. Evidently competition among English lock manufacturers was very severe in the beginning of the past century but of all the patented lock constructions the Chubb lever design was by far the most efficient and successful, so that in later years its principles have been sustained in all better class lever locks, not only in England but throughout the world.

The original designs of Chubb levers are shown in Fig. 198. These consisted of a brass strip or plate and formed the bolt checking elements in the lock case. The levers swing on the pivot (a) riveted to the lock case and are pushed down by spring action. The bolt tail, sliding along the back plate of the lock case, is provided with a stump, which passes through the pockets (see Fig. 199) and can move in these pockets in the horizontal direction. When turning the key the levers are lifted by the key bit until the gating is in line with the bolt stump, which can then pass through this gating to the next pocket when further turning of the key moves the bolt in or out.

The degree of security was determined by the precision with which the bolt stump passed through the gating and the pocket. In the same way variety in shapes and the number of lever steps in the key bit were determinants for the grade of such required. Finally the greater the number of levers in a lock, the greater the variety in the wards or steps in the key bit, as will be explained in the following chapters. In the beginning levers in a lock were of a uniform size and shape, the only variation being that the gating (Fig. 198 – c) in the lever could be arranged differently in a horizontal position. However, this lever construction underwent considerable improvements, as will be seen later.

The Chubb Detector

In this design the lever nearest to the lock bolt functions as the detector and is provided with a catch, checking the movements of the lock bolt. In addition a cylindrical pin riveted to this detector prevents overlifting of the other levers in the lock. In the event of attempts to pick such a lock by means of a false key or picking tools, the catch in the detector lever served to block the lock bolt, which could not be released until the proper key was turned in the direction to shoot the bolt, by which operation the detector by spring action returned to its resting position, and the functioning of the lock was restored to normal. A further security device, called the curtain, was incorporated in the lock. This curtain released the bellies of the levers when the proper key was used to shoot the bolt, but fully covered these bellies when the key was withdrawn.

Since the original invention of the lever principle, the shapes of such levers have undergone considerable changes, for the main task of such levers consisted in increasing and consolidating the protective power of locks against picking.

The illustrations on lock constructions given in Figs. 199 and 200 will disclose to the reader the great variety in lever designs.

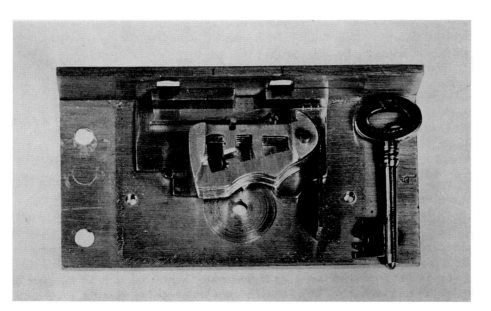

199. Chubb detector Lock and Key.

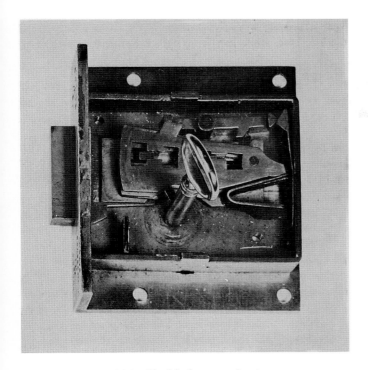

200. Chubb detector Lock.

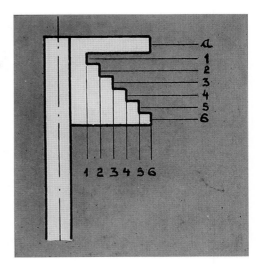

201. Principle of six differently sized steps in the key bit.

Levers

So far the author has not been able to trace the origin of the Dutch name *"Klavier"*. The Latin word *"clavis"* means key, whereas the word "lever" is used to characterize the security embodied in the mechanism of a key operated lock. In old literature the name of *"leuvel"* or *"leuvelveer"*, or the French word *"levier"* and *"gorge"* was found. "Lever" is the English name given to this lock part, whilst in the U.S.A. the word "tumbler" is found by the side of "lever". However the precise meaning—in the writer's opinion—is best reflected by the German word *"Zuhaltung"* (keep locked).

"Leuvel" is derived irom the French word *"Levier"* but is never used and represents vulgar Walloon introduced into the Netherlands by lockmakers from Liege (Wallonia) in the 17th century, where Liege lockmakers were highly reputed for their achievements in this field. In order to obtain a clear picture of the movements of a modern lever throughout its different stages when turning the key, the reader will undoubtedly appreciate a fundamental outline, illustrated by some sketches.

The six sketches in Fig. 202 represent six levers of exactly the same shape and dimensions, matching six differently sized steps in the key bit, as shown in Fig. 201.

By way of example a single throw lock containing six spring loaded levers is analysed here. Its corresponding key has six different steps, of which step (a) serves to shoot or withdraw the bolt and for this purpose engages the talon in the bolt tail. The black rectangular spots in the lever pockets indicate the position of the bolt stumps in these levers. These bolt stumps riveted to the bolt tail have to pass through their gatings with great precision, and so determine the security of the lock.

115

The curved edges, forming the bellies, in the lower part of the levers as shown in the illustration are all differently sized, and consequently the heights of the gatings in the levers have to differ accordingly, so that all in all the levers had two marked constructional deviations in common, viz: the sizes of the lever bellies and the heights to which the levers were to be lifted to have the bolt stumps pass the gatings.

The features of this construction are briefly reviewed here. When turning a key in a lock to move its bolt, the key bit will touch the curved edge of the belly in the lever. By turning the key further the lever is lifted by the corresponding step in the key bit until the bolt stump is exactly in line with the gating between the two pockets. Then key step (a) will engage the talon in the lower edge of the bolt tail and will push the bolt in or out. By these operations the bolt stump will be shifted from one pocket of the lever via the gating to the other pocket. After the bolt has been fully thrown or withdrawn, and consequently travelled its maximum distance in either direction, the levers by spring action will fall to rest on the bolt stump.

The question arises: Why should the sizes of the bellies of the levers be different? If they were not, as in the case of the original Chubb lever, the turning operation of the key would cause the levers to be constantly moving either upwards or downwards.

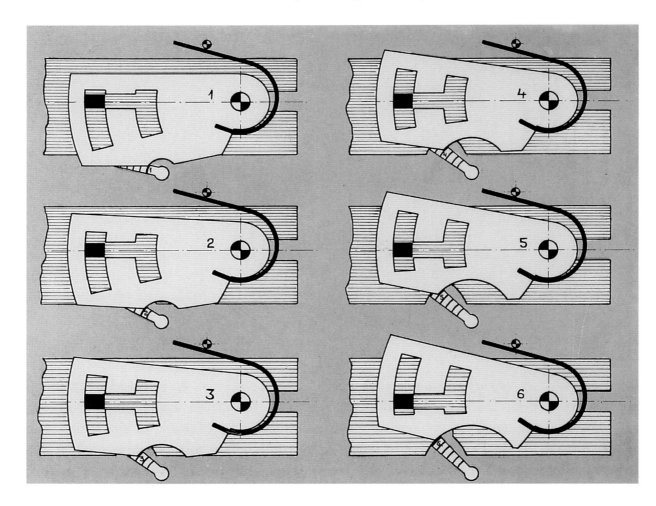

202. *A fundamental outline of six levers.*

As already said, a precise construction of the gating perfectly matching the passing bolt stump determines the degree of security of a lock. Now supposing the bellies of the levers were all the same size, the gatings unavoidably would have to be enlarged to compensate the vertical movements of the levers whilst turning the key. The greater tolerance in the gating would not only fatally impair the protective power of the lock, but would also minimize the variety of key constructions, since key steps, though differing in size, would allow the bolt stump to pass through these wider gatings. Summarizing the requisites for an ideal security lever lock, the writer finds:

1) Minute accuracy with regard to the thickness of the bolt stump, and the sharpness of its edges. Indentations in one of the faces of the bolt stump to correspond to identical tooth like notches in the pocket, as illustrated in Figs. 203 and 204, will mean a further asset to security of the lock.

2) The gating should match the bolt stump perfectly.

3) Precise construction of the curved edges of the lever bellies, to correspond to the steps in the key bit and their radii. Fig. 204 represents a Lips' lever, with lever belly corresponding to key step. Lever swings on pivot (a) bolt stump (b) has been riveted to the bolt tail; pocket (c) shows the tooth like notches; (d) represents the pocket right of the gating; (e) shows the lever resting on the bolt stump (b), with the bolt in unlocked position and (f) gives a cross section of the key and its steps.

However, not all locks marketed as lever or security locks are worthy of that name, in fact, only a very small percentage can be classified as real security locks, and then very often only up to a certain standard.

This statement is confirmed by the many burglaries easily carried out with the aid of false keys and lock picks. In this respect the writer fully endorses Mr. Price's statement that picking a real security lock is not an easy matter, but there is a great difference between the man in a laboratory testing with his own instruments how far a lock resists picking and the "professional" who under quite different circumstances puts the lock to a practical test.

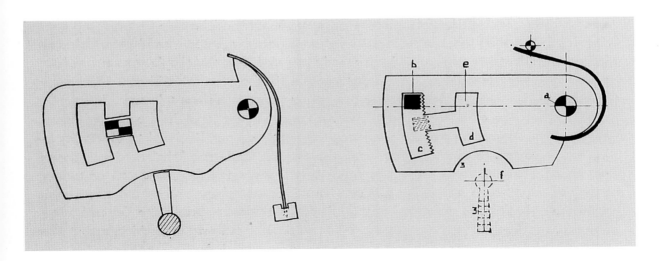

203. Principle English Chubb Lever.   *204. Explanation Chubb Lever.*

Security of Locks

A curtain is often provided as an additional security device, which covers the keyhole or the lever mechanism and which allows only the corresponding key to enter the keyhole and to operate the lock. The higher the precision with which the movable parts of a lock mechanism have been manufactured, the greater the accuracy with which the levers will pass through their gatings, a criterion for quality locks. The writer feels confident that his colleagues agree with Price's views on this score.

Discrimination and proper sense of judgment are indispensable factors for buying a real security lock, and it pays to select a guaranteed product of a reputable manufacturer whose sense of responsibility towards the community ensures the marketing of products worthy of his name and trademark. Although the question of price is a point for consideration it should not be a deciding factor, and in this connection the proverb "Penny wise and Pound foolish" should apply. It has already been remarked that scores of lever locks are quite unworthy of the qualification, but severe competition amongst manufacturers has unfortunately played an important part in deciding the final product, while price cutting has led to savings in labour and wages, inevitably resulting in sacrifices in quality. Many less scrupulous lockmakers were tempted to incorporate in their locks circular bolt stumps, simplifying the assembling process, instead of the square or rectangular types, not to mention the finish of the gatings, which showed too great tolerances, condemning a lock right from the start. Many a guileless buyer, unconscious of his deed, is now a user of such inferior locks. On the other hand discretion must be used in the final destination of a lock, for instance, not all users can afford to buy expensive precision locks, and indeed, in many cases such expense would be entirely unwarranted. For instance, a lock to be fitted to a kitchen cupboard differs from the requirements for a lock for a room door, a front door or a garage door, which claim varying degrees of security, whilst locks for strong rooms and safes call for first class qualities.

In their design certain lockmakers aim at a minimum inside depth to their locks between the case and the cap in order to accommodate the keyhole as close as possible to the back plate of the lock. For certain purposes the overall depth of a lock plays an important part, and in many circumstances very thin lock cases are the only possible solution. To achieve this many lockmakers fit the levers wrong side foremost which, in the writer's opinion, is fatal to the security of the lock. After analysing the working principles of the Chubb lever, which is reputed for its high security, every reader will be able to grasp the great drawback connected with the reversed lever, as its security features and its manifold varieties in construction have thus been greatly reduced for the simple reason that the gating allows the bolt stump to pass from pocket (b) to pocket (e), without any accuracy, whereas the bolt should slide through this gating with a minimum tolerance. In order to avoid obstructions or interruptions in the movements of the locks the use of cone shaped or cylindrical bolt stumps constitutes an unwelcome compromise. Real makers of security reject such compromises since the position of the pivot has been wrongly balanced in respect of the gating, in other words a properly synchronized function of the gating and the turning points of the levers is out of the question. In those cases where locks of minimum inside depths are required, in order to bring the keyholes as close as possible to the back plate, levers of the vertical type are preferred. Levers of this kind may be fitted in such a way that they slide simultaneously and synchronously with the bolt.

Cylinder Locks

By the side of the lever lock the cylinder lock enters into the foreground. Its key operated mechanism is not contained in an ordinary lock case but is accommodated in a separate cylindrically shaped housing, in its turn incorporated in a suitable casing. The fixing of a rimlock is done by screwing the cylinder to the flat surface of a door, whereas a rim mortice lock is fixed into a door with screws into a lock case, recessed in the door.

It has already been mentioned in an earlier chapter of this book that when designing his pin tumbler mechanism, Linus Yale was inspired by the ancient Egyptian peg system. His aim was (1) improved security against picking; (2) a handy small sized key; (3) a widespread use of the system.

In achieving his aim Yale brought about a revolution in lock manufacture, which at that time had already greatly developed, and had made considerable progress. There is hardly any lock that has undergone so many changes as the cylinder lock.

Many patents have been granted, some of which will be discussed and illustrated.

The pin tumbler mechanism is contained in an outer cylinder, in which an inner cylinder is rotated by turning the key. Three, four, five and even six cylindrical holes, drilled in the outer cylinder contain a corresponding number of spring loaded pins, each in two parts, which are to be operated by the various cuts in the key. The pins differ in length and have to be lifted by the key until the division line of the pins coincides with the line between the plug and the cylinder. The plug, to which is attached a connecting bar or tongue to operate the bolt, can then be turned by the key, by which operation the bolt is released and can be moved in or out.

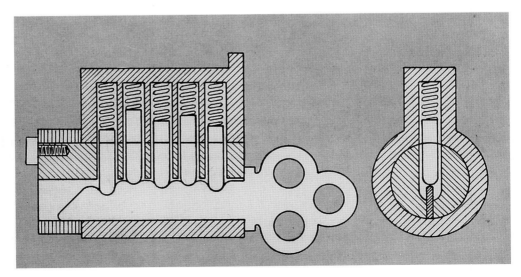

205. First five-pin cylinder lock from Yale.

A five pin cylinder lock already allows of a great variety of combinations or permutations. When the small slots in the circumference of the pins are made to differ only 0.5 mm in sequence, this would yield ten new combinations or permutations in the corresponding key bits and finally result in more than 100,000 variations in the key steps. In the same way a six pin cylinder lock would produce over 500,000 combination possibilities. However, these mathematical considerations are of a purely theoretical nature, since variations in depth of the key steps are necessarily limited in order not to weaken the rigidity of the key shank by too deep cuts. For these practical reasons the number of key step variations should not surpass 25,000. When Yale marketed his first cylinder locks, they were supplied with flat keys made from sheet metal, as shown in Fig. 205. The flat key soon turned out to be lacking in rigidity and a poor match for the properly shaped key guide in the inner cylinder of the lock. The cuts or steps often claimed half the width of the key shank, owing to which such a key after having been inserted in the lock could be moved up and down; this inadmissible tolerance caused incorrect lifting of the pins. A better guidance was ensured by the use of a corrugated key, a sheet metal key with a groove or grooves, matching the tongue attached to the inner cylinder. Illustrations Nos. 2, 3 and 4 in Fig. 206 show keyholes corresponding to such grooved flat keys.

However, the struggle between lockmakers and burglars has continued and both parties have always been on the alert watching each other's movements. By means of lock picks of every design, mostly made from flat spring steel, burglars succeeded in adjusting the pins to suit the division line and opened cylinder locks. Lockmakers again outwitted thieves by providing their locks with zigzag corrugated keyholes (sub 4) to prevent a flat strip from entering. Although corrugated keyholes mean a considerable improvement over previous systems, yet experience showed that they did not offer the ideal protection against picking. Eventually the paracentric key guide proved a worthy opponent to unauthorized intruders. The top part Fig. 206 clearly shows the principle of the paracentric key guide; three bullets or notches a, b and c overlap the vertical axis of the keyhole, which makes it impossible for a flat strip to be inserted.

The materials used for the construction of the component parts of locks are alloys of two or more metals in certain proportions, of a composition capable of resisting atmospheric and other influences, which might otherwise cause oxidation and would ultimately shorten the life of the lock considerably, not to speak of the defective functioning of the mechanism during its lifetime. For all these reasons the cylinders are made of bronze, the pins of a hardened nickel and bronze alloy and the helical springs of phosphor bronze.

On his way to the U.S.A. in 1903, when the Dutch Lips' Lock Factory was still only a small workshop, the writer comfortably seated in his deckchair was looking over his literature on cylinder locks, when it suddenly came to him that practically any cylinder lock, as then actually used, could be opened in a few minutes without much difficulty, as the component parts of cylinder locks consisted of alloys easily accessible to a drill. As already explained, the inner cylinder is made to rotate in the outer cylinder by placing the slots in the pins in alignment. Now the only thing to be done was to drill a 3 mm. hole just above the keyhole. In this way the pin tumblers were radially cut into two, and thus all obstructions were eliminated.

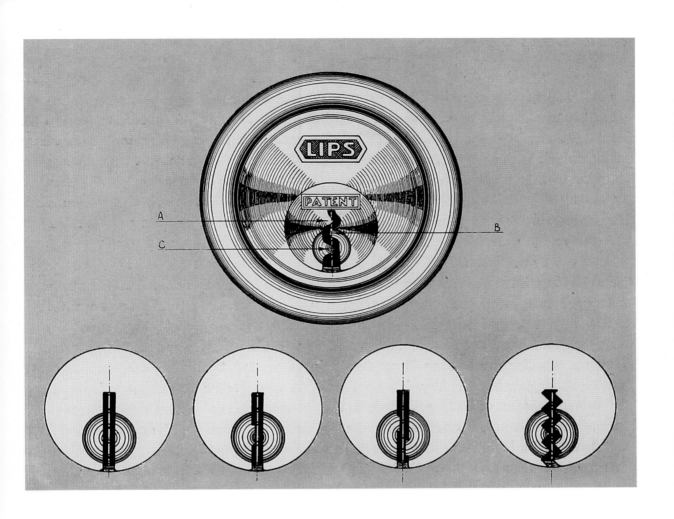

206 Different Key holes in plug.

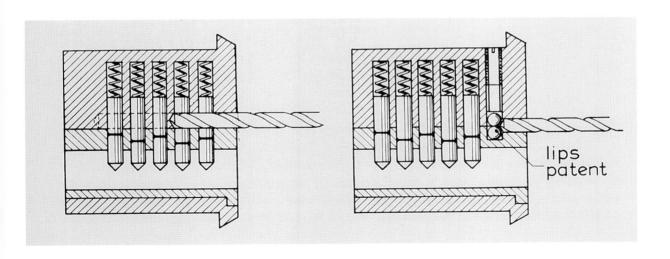

207. Cylinder lock, easily accessible to drills.

208. Lips' patented cylinder lock inaccessible to drills.

The lock could now be easily operated by a flat strip or blade. See Fig. 207.

This discovery was naturally uppermost in my mind until I had the opportunity to discuss it with Mr. John Mossman of New York, well known American lock expert, who as a bank architect designed and controlled safes and bank vaults throughout the U.S.A. When the writer told him of his discovery, Mr. Mossman was amazed and immediately ordered severe tests to be applied along the lines indicated. The outcome of the experiments confirmed the writer's views but added to Mr. Mossman's uneasiness and anxiety as thousands of safe deposits in the U.S.A. were fitted with cylinder locks of the type under discussion. The writer was able to suggest some solutions to remedy the situation, one of which proved highly valuable and efficient and for which a patent was subsequently granted. This solution was very simple indeed. The fundamental idea was to make pin cutting by drilling virtually impossible, which was easily achieved by shifting all the pin tumblers about 4 mm. further back in the cylinder, thus releasing the hole of the front pin tumbler. In this front pin hole a hardened-steel drill-resisting screw was fitted with two hardened-steel balls vertically placed just below this screw and likewise resisting any drilling effort. The lower ball was recessed in the inner cylinder. This device could be fitted to all existing cylinder locks at a minimum cost (See Fig. 208).

The reason for dwelling on this subject is to bring to the foreground the enormous scope of this invention, which laid the foundation for the outstanding success of the Lips' cylinder lock, regarded as a symbol of security. Several more patents were applied for and granted covering a variety of technical improvements in the Lips' cylinder locks, so that the quality of the final product rose high above any other make.

A striking illustration of the number of possible combinations in cylinder locks was afforded by negotiations with a foreign shipping company for the supply of 5,000 cylinder locks to be distributed over ten decks of one of their big liners. It was conditional that all locks should be different. The chief steward was to have at his disposal a submaster key fitting all the locks on his deck, whilst the captain was to have in his possession the grand master key controlling the whole range of locks in the ship. This transaction is not different from the normal requirements of locks divided into two or more different suites for large buildings. However, specific demands have to be met when a big mail steamer is concerned. A passenger may bolt his door on the inside, which prevents it from being opened from the outside, but supposing a passenger is seasick and wishes to call in the doctor, he may be too sick to leave his bed, thus the doctor has to consult his patient from outside the bolted door.

The chief inspector of the shipping company intimated to me that manufacturers in his country could not satisfy his demands, which they considered impracticable, so that he was obliged to find another source of supply.

The Lips' laboratories then set to work to find a scheme to enable the ship's doctor to open the bolted door by means of a special key to fit the normal keyhole. This led to the construction of a special key, which permitted only the ship's doctor to unbolt the cabin door and treat the patient. Neither the captain's master key nor the steward's submaster key could release the bolt. On the other hand the doctor's key could not operate the bolt if the passenger, the chief steward or the captain had thrown the lock bolt with the keys in their custody. Thanks to this special device an important order was passed to the

Lips' factories to supply the locks for that particular steamer. However, another feature was added. In case the passenger had to be sent to the ship's hospital for some days, the doctor could enter the patient's cabin at will during his absence. The deck steward may rightly object to the doctor's facilities, since he is held responsible for the passenger's luggage, and to eliminate this obstacle the Lips' factory incorporated a special device in the lock enabling the steward to lock the cabin in the absence of the passenger, but making the doctor's key inoperative. Consequently, due to these special provisions and precautions each holder of his particular key could make use of it as required for his specific duties.

Lock Picking

The writer must of necessity refrain from going into details on this subject. From a psychological point of view, making such practical knowledge public would not be a wise policy, besides any such information on the practice of lock picking is only of interest to the initiated craftsman, who is bound to benefit by such knowledge in his endeavours to design pick proof locking devices. Thus censure on moral grounds is imposed on any writer. This side of lock technique demands an equal amount of skill and practice required for the making of high grade locks, since only a thorough insight into the methods applied in lock picking will enable the lock manufacturer to build pick proof locks, guided by these subtile and ingenious methods. Generally speaking, the mere sight of a keyhole construction should enable the expert to determine to what extent the lock mechanism resists picking attempts.

Obviously a lock worthy of the name of security lock should withstand picking and be only accessible to the proper key. If not so, the lock mechanism lacks precision or refinement.

209. Mr. Hobbs, an American lock expert, in The Times of Sept. 4th 1851.

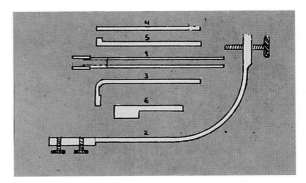

210. Instruments Mr. Hobbs used in this particular case.

It is to be regretted that a lot of locks in circulation should be classified as unsafe and considered 'foul play' from the side of the manufacturers. Nevertheless, the lockmaking industry in general can be proud of its achievements. The question of the doors and doorposts to which security locks have to be fitted is, however, quite another matter and beyond the control of the lockmaker, even though such circumstances should not be entirely neglected.

In expert circles it has always been a criterion that any key operated lock is subject to picking, although picking a real security lock is not at all an easy matter as one is led to believe from detective and film stories. Naturally any security lock subjected to picking in the laboratory with ample tools to hand and no time limit imposed will yield, although the same lock will stubbornly resist the attempts made by the burglar to pick it in darkness and in addition harassed by the fear of being caught redhanded. Many of our present day locks can, thanks to the extensive research made in the laboratory, withstand picking. In the middle of the 19th century inventors in the lockmaking trade subjected their products to a challenge.

Joseph Bramah (England) displayed in his shop window a lock bearing the sign:
"Whoever can produce a tool to operate this lock without its proper key will be paid the sum of 200 guineas on the spot".

In 1832 Chubb of Wolverhampton challenged an expert lockmaker to force a lock of his design within a stipulated time. Mr. Thomas Hart took up the challenge, which was announced by the town crier, but could not finish the job within the time limit specified. Mr. Hobbs, an American lock expert was more successful in 1851, and opened the Bramah lock in 25 minutes, and was paid the sum of 200 guineas by Mr. Bramah. The illustration of Fig. 209 is a reproduction from The Times of September 4th 1851, showing Mr. Hobbs intensely engrossed in his task and Fig. 210 shows the instruments he used in this particular case.

Mr. Hobbs started by inserting a metal strip into the keyhole, to exercise pressure on the lock bolt, so that the bolt stump was pressed to the notches of the lever gatings. By the insertion of a second strip the levers were carefully lifted so as to allow the bolt stump to pass through the gatings and the bolt to be moved. Experience teaches best also in the case of lock picking, so that lock manufacturers were forced to do everything possible to improve their security locks and to use their ingenuity to overcome and neutralize the methods of picking then applied.

Lock picking itself demands an expert knowledge and skill based on a thorough study of the craft and a considerable amount of practice, patience, and perseverance. The successes of the lock pickers have always been a constant impetus to an improved construction, since these served to expose the weaknesses and technical deficiencies of supposed security locks.

It is a well known fact that a lockmaker while boasting of the integrity of his products will at the same time make lock picking one of his hobbies. He will sit for hours experimenting with, and testing the weak and strong points of his lock constructions and will turn his hand to designing the most efficient tools, thus lockmakers may rightly be considered lock picking experts, although in many cases the picking of a lock is a matter of sheer luck. Needless to mention, this systematic studying of lock picking and the required tools will enlarge the lockmaker's knowledge and will inevitably lead to improvements in his security technique. The author has also spent many an hour on such study and has built up an interesting collection of picking tools by the side of his comprehensive collection of locks and keys. The photograph on this page shows him in the act of picking a lock and an illustration of a series of picking tools can be glimpsed in Fig. 212.

211. *The author in the act of picking a lock.*

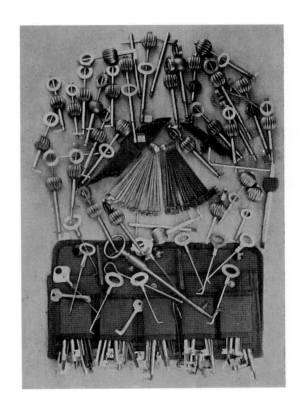

212. *A series of picking tools.*

CHAPTER VI
Permutations in Locks and Keys

Undoubtedly, the reader will appreciate some detailed information on the ways in which permutations and combinations, i.e. variations in lock and key arrangements, are obtained. By permutations we understand the particular successive order in the possible variations, whereas in combinations no such order in the arrangement of letters and ciphers is considered. In the latter instance such arrangement is arbitrary in most cases. Permutation refers to key operated locks, combination to letter and cipher locks.

An analysis of the lever principle has led to the conclusion that the shapes and depths of the key steps and the shape of the lever bellies in the lock should match precisely. On the subject of lever locks it should be observed that the number of exactly fitting levers in a lock predetermine the number of possible variations and so do the number and degrees of "liftings" dependent on the shape of the levers. Finally, the precision with which the bolt stump fits in and passes through the gatings counts heavily here.

For a really good lock the key is made first, which is the one proper procedure in lock making and allows of the variations under discussion. The order of the variations in the key bit should be systematically arranged and laid down beforehand on a so-called permutation sheet. Each step in the key bit is classified in these permutation sheets by a special number and all the steps combined are indicated, too, by a series of consecutive numbers corresponding to the range of levers in a lock. Thus a permutation sheet constitutes a systematic arrangement of figures, each of which represents a step in the key bit.

To illustrate the result of a very simple permutation Mr Price mentions in his book the following example:

The product obtained by a simple multiplication of the number of times 12 bells can be rung or sounded in mutually different order will be $1 \times 2 \times 3 \times 4 \times 5 \times 6 \times 7 \times 8 \times 9 \times 10 \times 11 \times 12 = 479,001,600$. In the same way multiplying the mutual changes of the order of the 26 letters of the alphabet will yield 403,291,461,126,605,635,584,000,000 as the product. It will not be difficult to realize the astronomical figure showing the number of variations in the sizes and shapes of steps in the key bit designed to operate a 12 or 24 lever lock.

Figures like these will surprise those uninitiated in permutation work, and many suddenly faced with such a problem, based on every day life, will scratch his head, shrug his shoulders and leave the problem unsolved. Let the reader judge for himself from the following anecdote:

One day a man, wishing to dine with a family of 6, suddenly dropped in and asked his host to tell him straight away the amount he would owe him supposing he was to pay for the dinners, if the members of the family and he himself would change their places every day. His host unaware of the purport of the simple question quoted a trifling amount for which his guest could join him and his family on 5,040 consecutive days. He thanked the host for his hospitality.

The key reproduced in Fig. 213 in addition to step (a) shows 6 further steps (1-6). Step (a) serves to move the bolt in or out, the steps 1-6 operate the levers of a six lever lock.

Supposing each of the levers is lifted to its proper height, determined by the steps of the key, which cause the degrees of lifting to be mutually different, the number of permutations in the steps of the key bit is obvious, for how many numbers of 6 figures can be composed, if each figure can be arranged in different order, to make up a different number each time.

Theoretically speaking, a shifting of figures may show the following series:

111111	111121	111131 etc. up to incl.	666661
111112	111122	111132	666662
111113	111123	111133	666663
111114	111124	111134	666664
111115	111125	111135	666665
111116	111126	111136	666666

From a technical point of view this way of shifting figures would result in a considerable amount of impracticable key steps, since the regular consecutive order of numbers e.g. 222222 — 222223 — 222333 —-111332 — 555566— 666666 — etc. would produce a range of highly inconvenient, or rather useless key steps. All this holds good for keys of lever locks, for cylinder locks this way of shifting figures is not objectionable. Keys of cylinder locks allow of such arrangement on account of notches between the intervals. In view of this the consecutive order of figures is interrupted by skipping a number of such arrangements, so that for lever locks highly practicable permutations are obtained viz. the product of $5^5 \times 6 = 18,750$.

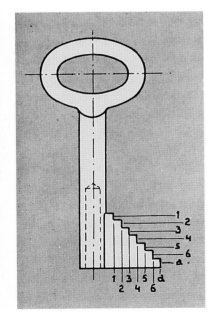

213. Key in addition to step (a) shows six further steps (1-6).

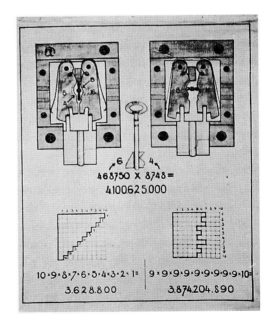

214. Example of a simple multiplication of the number of variations.

Many lever locks are offered, the levers of which differ in number, size and shape. Double bitted keys allow of a double lever arrangement running up to 10, 12 or even 16 levers. Besides, locks with a double set of levers are operated by double bitted keys, of which the bits differ in size to prevent the key from being wrongly inserted in the keyhole.

Further, each key bit of such keys has its own specific permutations, and by multiplying the resultant permutations of the two key bits the final amount of the possible permutations in such keys can be easily concluded. The illustration in Fig. 214 may serve as an example of such a double bitted key and its range of permutations.

Every reputable lock manufacturer is bound to maintain permutation sheets, but it is to be regretted that not all manufacturers fulfil the requirement by consistently and systematically keeping such records. Only manufacturers of real security locks can boast of an extensive well planned permutation administration. This is by no means a simple proposition, for compiling such permutation registers, maintaining and efficiently applying them is an exacting task.

Assuming one man could work out 10 permutations every minute, to work out a 12 lever lock would take him 91 years, 3 weeks, 5 days and 5 hours. It again holds good here that not all locks presented as such are in reality security locks. The general tendency to buy cheaper and still cheaper must inevitably favourably influence the sale of so-called low priced locks. However, users will soon learn to their cost that in the long run the more expensive article is the best and the cheapest.

Combinations

Combinations are used for keyless letter and cipher locks, which are operated by placing certain numbers of letters in a particular order. Combinations allow of numerous variations since the order of the numbers or letters can be changed at the desire of the lock owner. A letter lock, operated by a rotating dial with a 24 letter scale, affords $1 \times 24 = 24$ combinations. In the case of two or more dials the number of letter combinations can be extended, viz. for a double dial lock $24 \times 24 = 576$, for a three dial lock $24 \times 24 \times 24 = 13,824$ and for a four dial lock $24 \times 24 \times 24 \times 24 = 331,776$ combinations. A rotating dial with a 48 letter scale on the outer knob, split into 24 capital and 24 small letters, will produce $2 \times 24 =$ in case of a single dial, $2^2 \times 576$ for a two dial, $2^3 \times 13,824$ for a three dial, and $2^4 \times 331,776 = 5,308,016$ combinations for a four dial lock.

Modern cipher combination locks usually contain dials, each of which is provided with a numerical scale of 100 numbers, resulting in the following numbers of possible combinations: for a single dial lock 100, for a double dial lock $100 \times 100 = 10,000$; for a three dial lock $100 \times 100 \times 100 = 1,000,000$ and for a four dial lock $100 \times 100 \times 100 \times 100 = 100,000,000$ practical combinations.

A full description of cipher combination locks will be given later.

Modern Locks and their Uses

It stands to reason that it is impossible within the scope of this book to discuss the entire range of locks and locking devices applied nowadays. We have to content ourselves with a survey of such locks as are within everybody's reach and applied in dwellings, offices, factories, in public buildings etc. Descriptive literature published by several lock manufacturers offers such an extensive range of locks which enables prospective buyers to select locking devices to suit their particular and varied requirements.

The author purposely refrains from showing preference or bias for any particular makes, his sole object being to help readers to a proper sense of discrimination; nevertheless the reader will undoubtedly make allowances for the writer's prejudice in favour of the Lips' achievements in the field of locks, to which he has contributed his share in his 58 years' activities with the concern.

For use in houses different kinds of locks are utilized, mainly those for the doors of rooms, cupboards, wardrobes, sliding doors, and last but by no means least the front door.

Now let us first discuss the requirements for cupboard and room locks. More than 55 years ago, when the writer started his career in the lockmaking trade, the quality and security of cupboard and room locks was far inferior to those employed nowadays for modern houses and buildings.

Cheap and low grade locking devices of all kinds of origin were readily sold and some such locks were known as the *"Hollander Schlosser"* (Dutch locks), which was by no means a compliment. Even at that time lock designing and manufacture were on a much higher level in Germany, England and particularly in the U.S.A. Since then lever and cylinder locks have been substituted for the old fashioned, poor quality, warded locks, in doors of rooms and cupboards. The use of these cheap grade warded locks was not confined to Holland only, but they found their way to Denmark, Belgium and also to South European countries and the U.S.A.

It was a long time before even England, the U.S.A. or Germany were capable of producing locks which could be classified as better quality locks for the purposes mentioned. When travelling through those countries, however, sturdy hand made locks could be seen, whereas the products destined for export were mainly inferior in quality and not worth the price charged.

Price competition between British, American and German lock manufacturers was highly detrimental to the quality, although credit must be given to those lock makers who persisted in supplying real security locks to the importing countries.

What requirements must locks for ordinary domestic use fulfil? It is obvious that a cupboard in a kitchen or a room is not the proper place to keep money or other valuables. Nevertheless a housewife has all sorts of things which she wishes to keep out of the way and for this she requires a locking device to provide such protection for her personal property, and the modern lever locks can be considered ample guarantee.

Locks for the doors of rooms are of the same character. In the majority of buildings all the locks in the different doors have different permutations, and the range of 24 - 48 - 96 etc. will prove a highly convenient permutation for the locks of an average building.

Scarcely any locks are so continually used as in the home, in particular those of room and front doors, so that locks for this purpose are bound to be of excellent quality, and can actually be found in modern lever and cylinder locks, now readily available at reasonable prices (Fig. 215).

Both for twin and single leaf doors the distance between the centre of the keyhole and the fore end of the lock should match the width of the door style and should leave ample space between the door frame and the keyhole, so that the key can be turned without injury to the user's hand by touching the door frame. In the same way the distance between the lever handle of a door in locked position and the door frame should likewise enable the user to keep clear of the door frame when operating both handle and key.

In European countries these distances usually vary between 40 and 60 mm for such upright locks.

As a rule, such correctly proportioned high quality locks have brass or bronze mechanisms, incorporated in steel lock cases; however, for better class buildings the brass or bronze construction is often required both for the mechanism and the visible outer parts of locks to avoid rusting.

In all cases a solid striking plate fixed to the door frame is essential in order to prevent the bolt from being easily disrupted. In practically all European countries the mortice lock is by far the most popular type. Apart from ships, rim locks are seldom used and steamship companies are more and more inclined towards the use of mortice locks. Door locks for rooms are mostly mortice, cupboard door locks can be either mortice or flush locks, and in some cases rim locks. The difference between the lock types mentioned is explained somewhat more fully, as follows:

1. A Mortice lock is a lock which is inserted in a hole cut in the edge of the door.
2. A Flush lock is recessed into the face of the wood of the door style, to lie flush with it.
3. A Rim lock is fixed on the face of the door style.

A great drawback to locks made of steel is that a painter can never leave them untouched, and he invariably includes the fore end and the bolthead in his job. A painted bolt cannot function properly, because the paint sticks to the bolthead and partly to the bolt itself. To avoid this inconvenience it is recommended to remove the lock from the door before the painter starts on his work and to replace them afterwards. This suggestion is made not only for cupboard door locks but indeed for all the locks in a dwelling house or other buildings. In several countries such as England, France, Germany and U.S.A. many doors are not painted at all and the natural colour of the wood is maintained. In such countries locks and furniture are not fitted by a carpenter but by experts of the lock industry, who confine themselves exclusively to lock fitting jobs. The proverb "shoemaker stick to your last" applies here, for a carpenter cannot be expected to be a lockmaker, and a lock maker cannot be expected to be a wood expert.

The policy of assigning these tasks to specialists is gradually being adopted in the Netherlands, too, and quite rightly so. Because of persistent use, locks for the doors of rooms should fulfil the highest technical requirements.

215. A Mortice lock for ordinary domestic use.

216. Lever handle lock construction.

217. Knob operated lock construction.

One moment's thought will make every one realize how frequently room or passage doors are opened and closed every day, not to mention the number of times a door slams due to draught or otherwise. It can safely be said that locks are supposed to last for a life time, but are also the least cared for. Overhaul and maintenance are usually neglected; in fact, locks are seldom touched with oil or grease, but nevertheless it is surprising to see for how many years such locks survive without falling into disuse. Two fundamentally different designs have to be considered for the locks of doors of rooms:

1. The lock fitted with a lever-handle which is pressed downwards to withdraw the spring bolt.
2. The lock fitted with a knob, which is to be turned to operate the spring bolt.

The difference in the designs of the two lock mechanisms is clearly shown in Figs. 216 and 217.

Nowadays the lever-handle lock is undoubtedly the most popular in Europe. The latch bolt of a lever-handle lock is operated by a lever handle on both sides of the door. The latch bolt is lightly spring loaded, so that it shoots out automatically as soon as the handle is released. Excessive spring pressure is not desirable as it results in door closing being more difficult. Further, since the latch bolt returns with a bang this can bring about mechanical fracture or breakage at the edges of the door style.

The latch bolt is moved by pressing the lever-handle downwards, which resumes its neutral, i.e. horizontal, position after the latch bolt has been moved out due to the spring action. The spring arrangement of the knob operated locks should be much lighter for easy turning of the knob and consequently the springing of lever-handle and knob operated locks is bound to be entirely different. A knob is hard to turn against an adequately sprung lock or lever-handle, which implies that knobs or lever-handles both sides are preferable, with the lock spring to suit.

In conclusion, the difference in springing between lever-handle and knob operated locks excludes interchanging, since the axial movement of a knob demands less springing than a lever-handle. Lock manufacturers have always been conscious of the springing requirements particular to lever-handle and knob operated locks, so that an expert will never risk interchanging them, whereas a carpenter is usually quite ignorant of these characteristics. The lower part of the lock case of a room door lock comprises the key operated night bolt and its mechanism. In Europe two types of this lock are used:

1. One type has the keyholes horizontally displaced, that is not on the same perpendicular, which prevents keyhole peeping.
2. The other type has the keyholes on the same perpendicular, this avoids guides too long for the key and the handles.

The writer has too often noticed the ignorance of many a teacher in technical schools about the "why and wherefore" of fixing locks in the prescribed way, nor do some teachers know the difference between a lock mounted with a knob or lever-handle. How then can a "skilled workman" be expected to do his job properly? In fact, technical training in this respect often leaves much to be desired, although credit must be given to those teachers in technical training schools who are truly conscious of their task. The writer, by the way, is fully prepared to supply any information to technical schools on the products of his branch of industry.

Front Door Locks

Space forbids detailing the various types of front door locks and their features, and only the most popular types will be discussed.
Obviously, a front door lock should be of a solid and sturdy construction and its requirements are entirely different in that respect from those of the usually lighter cupboard and room door locks.
Front door locks must also include a latch and a dead bolt and security is determined by the precision with which the levers and the bolt stumps have been designed and constructed, namely the accuracy with which the stumps pass through the lever gatings. During the day and in the evening the front door is normally secured by the latch or spring bolt by simply pushing or pulling the door. Consequently it is imperative that the latch bolt in itself can be relied upon for absolute security. The dead bolt is shot by the key for the night.
Here again severe competition and price fighting has seriously impaired the quality of existing front door locks, so that their latch bolts are far from secure, and easily opened; this accounts for the numerous thefts committed, particularly in larger towns. By means of a simple steel wire pick such inferior latch bolts are easily withdrawn.
It is not fair to the community, consisting mainly of laymen, to bring such rubbish on to the market, because persons who are not properly instructed on the merits of a good lock cannot be expected to appreciate such merits and their confidence in the products of the lockmaking industry is only too often abused.
In a front door lock both the latch and the dead bolt should offer equal security. The illustration in Fig. 218 (left) gives an inside view of the Lips' lock mechanism No. 507 for front doors. This lock, which is operated by a knob on the inside of the door, has two key holes and its latch bolt can be moved in, from the outside, both by the servant's *single* bearded and the owner's *double* bitted key. However, it is the owner of the double bitted key who can throw the dead bolt. Thanks to this design the single bearded key can safely be left in the custody of the personnel, since the owner can make this servant key inoperative by shooting the dead bolt with his double bearded key.
The illustration in Fig. 219 shows a very secure cylinder lock for front doors. The operation of the latch bolt in a cylinder lock is based on the lever or bar principle, by which the cylinder imparts its movement to the latch bolt. A second turn of the key will shoot the dead bolt and is kept in place by a heavily spring loaded lever or bar in the lock case.
Lately, separate cylinder units have become available to be built-in in numerous ordinary locks. In practice, however, such gadgets have proved quite ineffectual, since they are often incorporated in cheap locks of inferior quality, which are of too light a construction and failure is bound to result when used as front door locks.

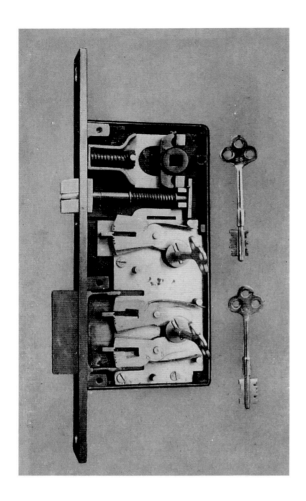

218. Inside view of Lips' Front Door Lock Mechanism.

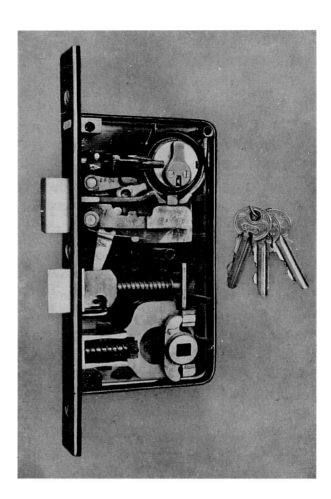

219. Inside view of Lips' Cylinder Lock Mechanism.

Master Keyed Locks

The master keyed system shows marked advantages when applied in offices, factories, hospitals and similar important buildings, hotels, prisons, flats and likewise on steamships, etc.

Master keyed locks are divided into two or more different suites, each suite comprising locks that can be operated by a number of different keys, but all of them controlled by a grand-master key. When locks are divided into two or more different sub suites, each sub suite can be controlled by a sub-master key which passes all the different locks of such a suite. However, the fact remains that the grand-master key, belonging to the General Manager, controls all the different locks of the suites. Furthermore the operation of the locks by the different keys and the sub-master keys can be neutralized by an extra turn of the manager's master key.

Master Keyed Lever Locks

Master keyed lever locks are not so widely used as master keyed cylinder locks. Master keying in lever locks can be arranged in two ways:

1. All the keys are inserted into one and the same keyhole.
2. The locks are provided with two keyholes, the one perpendicularly placed over the other. The master key is inserted in the top key hole, the sub-master and servant keys fitting in the lower.

In master keyed lever locks with a single keyhole, key variations are obtained by different wards and steps in the key bits or by studs riveted in some of the levers, so that one lever controls another. However, the great drawback to this system is that a servant key can easily be transformed into a master key, simply by filing the ward steps in the key bit to the required shapes and sizes. Some sketches in Fig. 221 clearly demonstrate how easily such changes can be made, when comparing the key bits A, B, C and D with the key bits A^1, B^1, C^1 and D^1.

Many such single keyhole master keyed lever locks are still being manufactured, but for reasons explained above their security is doubtful. Another disadvantage is that the key guides can be provided with brass strips properly grooved to match shape and size of the key pin. Such locks lack security as they are easily picked and are, therefore, to be rejected.

From the foregoing the reader will have concluded that such single keyhole master keyed lever locks are to be replaced by the double keyhole master keyed lever locks, as mentioned under 2, so as to solve the security problem.

It should be observed that the Lips' master keyed lever locks are provided with the double keyhole, which system affords convenient and secure master keying arrangements. Besides, the servant keys fitting the lower keyhole have a key pin of a larger diameter, so that it cannot be inserted by mistake in the top keyhole, which might cause damage to the lock mechanism. For this class of lock, no better system than the double keyhole has so far been found, as full security of the levers is maintained. For an example the reader is referred to the sketches in Fig. 222. In this construction D^2 can be the master key of A^2, B^2 and C^2, and so on.

The dead bolt in a double keyhole master keyed lever lock has two talons and is kept in the open or locked position by a series of levers operated by both keys. The levers in such locks do not rotate but slide. The cuts in the bit of the master key are entirely different from those of the servant key, so that necessarily the lever bellies have also to be differently shaped, to match the right key.

All keys, pertaining to this class of lock are of the pin type, which can operate the lock from either side. Suppose the manager enters a room, locks it on the inside and leaves the key in the lock, then no servant key will be able to operate the lock, because the levers are arrested by the manager's key. As soon as the master key has been removed the servant key will again function properly.

Fig. 220 shows an illustration of a six lever double keyhole lever-handle Lips' lock.

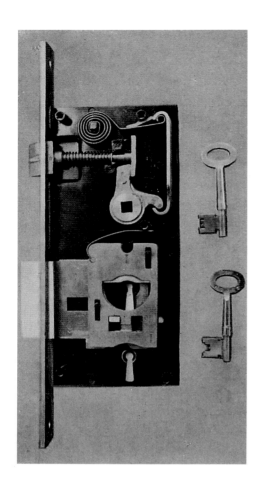

220. Lips' Master keyed lever lock with double keyhole.

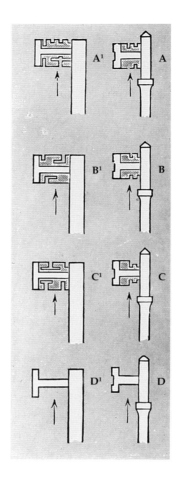

221. Master keys, for ward locks (D).

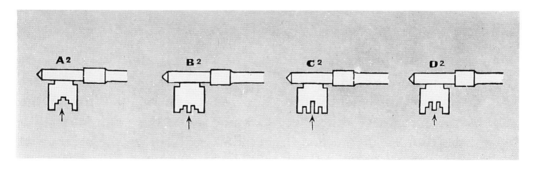

222. Master keys for lever locks (D^2) with two keyholes.

Master Keyed Cylinder Locks

The arrangements for master keying cylinder locks are much simpler than those for lever locks, for the pins in the pin tumbler mechanism are provided with notches to suit the various cuts in the key bit of the master key as well as the steps in the servant key, so that two different dividing lines of the pins are made to coincide with the line between the plug and the cylinder body, according to the key used, either the master or the servant key. Usually the master key has deep steps, whereas in case of a sub-master or a servant key the steps in the shank near the bow are applied.

Finally some general hints for the carpenter are given here in connection with a proper fixing of locks and their corresponding striking plates.

A properly fixed lock should work in the same easy and smooth manner as when tried in the hand. The main reason why many locks do not function as they should, is that the lock is too tight in the mortise, or in the case of a rim lock the surface of the door, when fixing the screws, was not sufficiently flat. The result will be that the slightest shrinkage or warping of the wood will distort the lock and cause the mechanism to become bound. Therefore, the mortise in the door should be cut slightly larger than the body of the lock for clearance in all directions, as shown in the drawing in Fig. 223 (A, B, C). It is quite satisfactory to have the fore end of a mortise lock carry the whole weight of the lock, if this fore end lies in a well fitting recess in the edge of the door (D). In the case of a rim lock a really smooth surface will prevent the lock from being strained when the fixing screws are tightened.

Striking plates are fitted in a recess in the door frame to receive the bevelled bolt. The striking plates are supplied with a bent lip, along which the bevel of the bolt slides when closing the door. On no account should the lip of the striking plate meet the bolt close to the fore end of the lock. The lip should be set or aligned by bending, so as to secure an easy strike, even at the cost of the architrave, which for this purpose might have to be cut away a little.

The first sketch in Fig. 223 shows a wrongly bent lip. The sketch in the centre shows a properly bent one.

Locks are invariably lubricated before leaving the factory, so that there is no need for further lubrication immediately after the carpenter has fixed the lock. Oil should never be used to lubricate cylinder or keys of cylinder locks. This should be done with a mixture of graphite or lubricant, which mixture will not become sticky.

Thanks to the co-operation of architects and building contractors, who have a keen eye for the requirements of good locks and their fixing methods, the principles as outlined above have been generally accepted in many countries, to the benefit of all parties concerned, viz. architects, contractors, lock manufacturers and...... the users of locks.

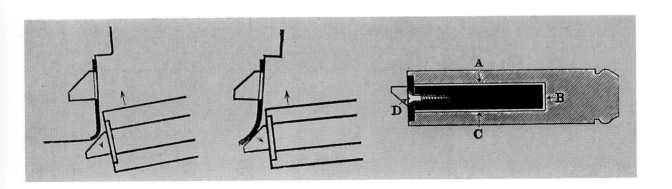

223. *A proper fixing of locks and their corresponding striking plate.*

CHAPTER VII
Locks for Safes and Strong Rooms

The reader will appreciate that it is impossible to mention and describe the hundreds of patents and constructions from the different countries, such as England, France, Germany, the U.S.A., Sweden, the Netherlands, etc., so that the writer is reluctantly obliged to make a rapid selection from the numerous and varied specimens of this class of locking devices.

Up to the end of last century, the key operated lock was practically the only type of lock known in Europe. Sometimes a letter combination device was incorporated in these key operated locks, which were then used in safe and strong room doors. For many years key operated locks of the Bramah type were applied on the Continent, both the Bramah lock proper and the combined Bramah-Chubb lock. In England, Price, Bramah, Chubb, Tann, Milner, Rattner, and many other experts were reputed for their security locks.

224 - 225. Bauche's Lock (Paris). Lips' collection.

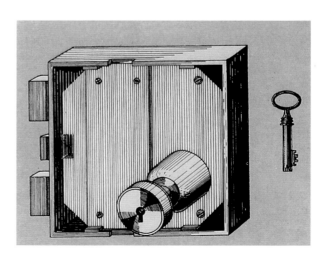

226. Disc dial.   227. A Bramah-Chubb system lock.

In France, Bauche, Fichet, Haffner; in Germany, Sommermeyer, Adé, Arnheim, Panzer; in Sweden, Rosengren and in the Netherlands Lips were the leading manufacturers of safe and strong room door locks. In the United States of America the chief makers were Yale, Russell and Erwin, Sargent, and Greenleaf, Mosler, Diebolt and others.

Although all these makes showed more or less important differences in design or construction, they had one thing in common, in that the greater part of the locks applied were of the rim lock type with a dead bolt, occasionally combined with espagnolette bolts fixed to the inside of the door, which bolts moved in four directions and engaged in the bolt holes of the frame of the safe.

The letter combination locks were of the three and four tumbler type of different constructions, whilst also the so called click locks were in use.

All the key operated locks mentioned had the shape of a cylinder, projecting from the lock plate of the case, containing the lock mechanism. Some European specimens of these cylindrically shaped locks are shown in Figs. 224, 225 and 226. These cylinders were mortised in the steel door, so that the fore end of the cylinder reached the front plate of the door, whilst the lock plate at the other end of the cylinder was screwed to the inside of the door, so that the cylinder was entirely recessed in the thickness of the steel door from front to back plate. An example of this method of fixing these cylinders is shown in Fig. 227, it illustrates a Bramah-Chubb lock.

This method of fixing these cylinders considerably reduced the security of such locks, since this security device was easily forced by burglars, unless it was protected by a special hardened steel plate.

Fig. 226 represents the disc and the dial of an old letter combination lock. The deeper tapering incision in the circumference works on the bolt mechanism, the three semi circular indentations are false notches. The spring loaded tumbler engaging the cams of the disc can be adjusted at will to change the letter combinations if so required.

The operating principles of the Bauche lock are clearly demonstrated in the illustrations in Fig. 224 and 225. By pushing the key into the cylinder the levers are set in accordance with the grooves in the key shank, so that the lever gatings can freely pass the notches in the ring.

Fig. 228 shows Fichet's click combination lock containing three discs. By inserting the key into the keyhole the three discs could be rotated, and by counting the correct predetermined number of clicks the lock could be opened. However, the great drawback to these click combination locks consisted in the fact that outsiders, too, could overhear the clicking arrangement of these locks. Another lock of French origin is the so-called repeater click lock, containing four discs, which could be rotated by a key with corresponding grooves. The word combination of this lock could be changed at will by inserting a lever key in the centre keyhole. Here also overhearing of the clicking arrangement by outsiders greatly decreased its security.

As already observed, key operated locks for safes and strong rooms were often combined with keyless combination locks to add to the security of the entire locking gear. To achieve a properly designed combination, a three or four dial letter lock was attached to the back plate of the key operated lock case, containing the bolt mechanism, and only after a predetermined letter combination on the dials had been brought into line with the setting mark could the bolt mechanism of the key operated lock be moved. In England mostly key operated locks were applied for safes and strong rooms, which locks were produced by the local industry. Their keys were of the single bit type, still widely used in Great Britain. Only in later years have some manufacturers switched to constructing locks for use with double bearded keys, Chatwood, Chubb and others.

228. Fichet's click combination Lock (Paris).   229. Repeater click Lock of French origin.

Kromer's Protector Lock

This key operated lock is outstanding in its class. It is a high quality lever lock, invented by Mr. Theodor Kromer of Freiburg-Breisgau and was protected in many countries by several patents, granted from 1869 to 1875, whilst many technical improvements on his lock were patented afterwards.

The writer, who knew Mr. Kromer very well, recognizes him as an inventor, designer and manufacturer of great genius.

Kromer's Protector lock was contained in a cylindrically shaped lock case, consisting of an outer and an inner cylinder. The inner cylinder, into which the sliding lever mechanism has been built, rotates. By turning the key, the sliding levers are set in such a way that some of the levers slide to the right and the remainder of them to the left, all in a horizontal direction, after which the inner cylinder is released and can rotate in the outer cylinder. The levers are spring loaded in order to make sure the levers will resume the locked position after the key has been withdrawn. According to Kromer the spiral reproduction of the "Protector" in Fig. 230 reveals the lock with the cap removed and with the key inserted and turned in the unlocked position. As may be seen in Fig. 231 of the "Protector" key, this double bitted key is a highly complicated one.

The bits are not only provided with the regular lever steps, but also with a number of irregular hollows and bevelled incisions, which in their turn work on special levers in the lock. For a minute setting of the levers the key bit contacted this mechanism at many different points during the locking and unlocking operation. The inventor made a real precision job of this and is firmly convinced that it is impossible to copy the key even when taking wax prints of it. The writer will neither deny nor confirm this statement. The writer many times copied a "Protector" key but from an original one, and found it a most delicate and exacting task, which can only be accomplished if the right tools and instruments for this particular key are available. After the expiry of the Kromer patents efforts were being made by German manufacturers to copy or improve upon the Kromer lock, but the writer has never found imitations that could equal an original Kromer, which was indeed a masterpiece of lockmaking. In fact, the writer has never heard of any successful picking of a Kromer lock.

Kromer's successors designed a key of which the key bit consisted of a number of rotary key steps, as shown in the illustrations in Figs. 232 and 233. After inserting the key into the keyhole the rotating steps are set or adjusted to match the levers in the lock, so that when turning the key, these steps could be set cross-wise in order to place the levers in the unlocking or locking position.

However ingenious this fundamental idea may be, the writer found that its materialization was attended with considerable mechanical difficulties and after all the security of the locking device was not improved since the number of possible permutations in the key steps was not increased. Finally the rotating key steps made the key a rather delicate construction. Fig. 233 shows the Kromer lock with its corresponding key, the picture on the left shows the position of the rotating steps, forming the key bit before the key has been turned in the lock, whilst the picture on the right represents the same key with the rotating steps set after the key has been turned.

230. Kromer's Protector lock (Germany).

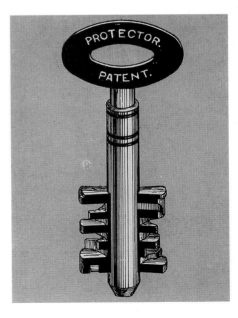

231. "Protector" double bearded key.

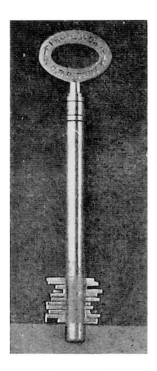

232. "Protector" Key with rotating key steps.

233. Rotating key steps differently arranged.

Criticism, if any, will obviously be limited to lock and key constructions used in the past. In as far as modern locking devices are concerned, the writer feels he should refrain from criticism and confine himself to a general description and analysis of existing lock and key constructions, with a view to imparting his practical knowledge to those interested. The lever mechanism of a key operated lock, based on the Chubb lever system, is considered even nowadays as a criterion of maximum security. Consequently, it is for this particular reason the Lips' factories have adhered to this approved system and have applied it widely in their various classes of locks (Figs. 234 and 235), adding many technical improvements.

The Lips' safe lock contains 16 precisely fitting and accurately balanced levers, the face of each being provided with notches or cams, and is operated by a double bitted key. The Lips' factories are proud of the fact that this type of lock has never been picked and may be considered entirely pick proof.

Precision is a first requisite in the manufacture of safe locks of such involved construction. For practical reasons the manufacture can no longer be considered bespoke, which would result in excessive cost and selling prices. In these modern times we cannot afford to turn out some stray masterpieces of craftsmanship, since the demand for safe locks has grown to such an extent that the manufacture in series of such locks is the only feasible solution to keep pace with this ever growing demand.

Such series production necessitated the foundation of factory buildings, equipped with special machinery and tools, the initial outlay for which could be compensated by a well planned annual manufacturing programme. Although, as a consequence of this mechanization of the lock making industry, the manufacture of parts and their assembly by hand have been entirely done away with, this does not reduce the craftsman to a mere robot, indeed he has every opportunity to display his initiative and skill in a modernized production plant. Only the craftsman realizes the requirements to be fulfilled to get proper planning in the production of modern security locks, of which the component parts forbid any tolerance in respect to dimensions and consequently their interchangeability. Indeed, mechanized production of locks and keys claims the expert's skill and cannot dispense with his experience, even though the technical aspect outweighs the artistic elements of his craft. After all, every key operated lock, whether destined for a safe or a strong room door, is bound to have a keyhole, passing through the entire thickness of the solid door and extending as far as the backplate of the lock case. The open keyholes have invited many a burglar to insert explosives and have facilitated many a burglary. Such keyholes offer excellent opportunity for depositing explosives and this method of opening locks was at one time being more and more frequently applied. It seems, however, that explosives are no longer popular, because nitro-glycerine endangers the burglar as much as the safe.

Nevertheless, safe manufacturers in the U.S.A. very soon realized the necessity of abandoning key locks for safes and strong rooms and introduced the so-called cipher combination locks, operated by a rotating dial, fitted to the outside face of the safe door. The original keyless combination lock was improved considerably and in the course of years its construction had reached such a standard of perfection that combination locks were being almost exclusively used for safe and strong room doors, thus excluding key locks for this purpose.

The first single dial combination lock, introduced into Europe and operated by a letter combination, chosen from two alphabets, was the Kromer letter lock. Its construction and method of operation differed fundamentally from those of the American combination, as will be shown later.

234. Lips' safe lock in open position.

235. Lips' safe lock in locked position.

During the past 50 to 100 years locks and their manufacturing processes have undergone considerable changes and improvements in the U.S.A. and Europe. The method of placing obstructions in or around the keyhole, to increase the security of a lock, was entirely abolished. Lock manufacturers gradually came to realize that real security could only be achieved by incorporating into the lock mechanism a number of movable (sliding and rotating) parts, each having its specific function in the mechanism, determined and directed by the steps in the key bit. Barron, Chubb and other designers may rightly be considered great pioneers in the field of modern lock construction and manufacture, while in the U.S.A. it was Yale, Sargent and Greenleaf, Diebold, Mosler and others, who have contributed most towards reforms in lock designs. About a century ago it was generally accepted that key operated strong room locks were pick proof. However, Dr. Andrew of Perth Amboy, New Jersey, was of a different opinion and constructed a key operated safe door lock of unusual design.

The main features of his lock were that the lock and key bit, which could be taken to pieces, comprised certain component parts which could be re-arranged and re-assembled at will to form a new permutation. Dr. Andrew, with this invention, wanted to prevent the key from being copied, which eventually proved a failure on account of the arbitrary arrangements of the permutations in the lock and key bit. This lock, which contained a series of tumblers (levers) and a detector, showed a considerable disadvantage, because for every new permutation, the lock had to be removed from the door and taken to pieces, altogether a rather complex manipulation.

Dr. Andrew's invention coincided with a similar invention by Mr. Newell, one of the managers of the firm of Day & Newell of New York. His lock construction, patents for which had been applied for, was fundamentally identical to Dr. Andrew's, the only difference being that a re-setting of the permutations in Mr. Newell's lock could be effected without having to remove the lock from the door or having to dismantle it. This facility placed Newell's lock on a much higher technical level. In order to obtain a new permutation, the detachable key steps composing the key bit were assembled to take up a new predetermined order. This being done, the "revised" key was inserted into the lock in the unlocked position and by simply turning the key, the lock mechanism was automatically adjusted or set to comply with the new arrangement of the key steps. It should be noted that Mr. Newell's levers consisted of two sections, one of which was lifted by the key bit, whereas the other was set by turning the key. When the key was turned half-way the second section would engage a number of indentures. A specimen of the Newell lock is in the Lips' collection and an inside view of the mechanism is given in the reproduction of Fig. 236. The keys, corresponding to this lock, were bulky and rather heavy.

The lever edge of the Day & Newell lock, reproduced in Fig. 237, shows a number of cuts or notches near the section marked (b). By turning the key halfway the levers were raised according to the steps in the key bit in their highest position. On a further turn of the key the bolt was moved and the set of levers adjusted to this new permutation. The tongues (a) were set by the cuts or notches (b) and the tumbler (c). Unlocking could only take place by using a key with steps, corresponding to these cuts or notches, so that the levers were released and the tongues (d) could slide into the opening (e).

236. A Day & Newell lock and the corresponding key. Lips' collection.

237. Inside view of a Day & Newell lock. Lips' collection.

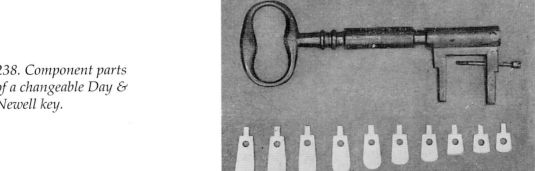

238. Component parts of a changeable Day & Newell key.

Obviously, any key with a differently arranged key bit would change the setting of the levers and would arrest or block the further movement of the lock mechanism. Fig. 238 contains a dismantled keybit, appertaining to the lock just described. The component parts of the key bit could be arbitrarily arranged to new permutations. A screw was passed through the holes in the key steps to fix them and to keep them in place.

Special mention must be made of the achievements of Mr. L. Yale and his son, who designed similar locks in different constructions. The Yale Double Treasury Lock, shown in Fig. 239 and 240, may be considered one of the finest locking devices and an outstanding example of supreme craftsmanship. This lock, of which there is a specimen in the Lips' collection, was designed and patented in 1852.

It combines volume and heavy weight, for it is 360 x 280 x 60 mm, in size and weighs over 25 kilos. As a double lock it has two keyholes and the double mechanism can be operated either by one or by two keys, each corresponding to one part of the double mechanism. Both mechanisms could be set according to the predetermined permutation or permutations of the key, which implies that the setting of one mechanism could be different from that of the other. The advantage of a double keyhole arrangement lay in the fact that the lock could be opened even when burglars had tried to pick the lock by force and had damaged one of the keyholes. The second keyhole was mounted entirely obscured from view and its secret position was known only to the manufacturer, who in an emergency could unlock the door via this second keyhole.

Permutations of both lock and key could be changed at will. The steps of the keybit were composed of detachable tongues, and thus could be easily carried about. On account of its complexity this lock defies detailed description, but it may be considered a real masterpiece of the lockmaking craft.

239. Yale's changeable key Bank Lock (locked). Lips' collection.

240. Yale's Double Treasury Lock, with bolt withdrawn.

In 1857 Linius Yale Jnr. designed and constructed his Double Quadruplex key lock, reproduced in Fig. 241. It is worthwhile giving a description of its composition and its principles of operation. Although its mechanism had only one fixed permutation, its high grade of security was obtained by the pin locking principle, inspired by the ancient Egyptian pin operated wooden locks. It has two cylindrically shaped keyholes and the corresponding keys are provided with four irregular vertical grooves in the key shank. Now each of these grooves is to set a series of five pins, so that altogether 20 pins are to be set before one of the keys can actually move the lock mechanism. The shapes of the two keyholes are different and so are the keys. Besides, the keyholes are entirely covered and blocked when the mechanism is in its locked position and access to the keyhole is regained only after a special double handled key has released the first keyhole. Subsequently by turning the normal key in this keyhole the second keyhole is released and made accessible. Finally, by inserting and turning the corresponding key in the second keyhole the bolt will be moved in or out. This total combination of lock and keys has the respectable weight of about 35 kilos.

Herring's Grasshopper Lock

This is a combined key and knob operated lock. Its small sized key, made of bronze, has a key bit, composed of five pins of different lengths. By inserting this key into the slot, which is partly cut out in the dial, and by subsequently turning the knob, the levers in the mechanism are adjusted by pushing action and the bolt is released. By a further turn of the knob the bolt was fully thrown or withdrawn, after which the key "hops" out of the slot, hence the name "Grasshopper" lock (Fig. 242).
Herring's Grasshopper "pin" key lock was designed by Herring of New York in 1848 and was considered one of the most secure locks of its time. A specimen is in the Lips' collection.

241. Yale's Double Quadruplex key lock. Lips' collection.

242. Herring's Grasshopper lock. Lips' collection.

Isham's Register Lock

Isham's invention was a kind of key combination lock. It was manufactured by the New Britain Bank Locking Company in 1858. The mechanism of this outstanding masterpiece of lockmaking craft, a specimen of which is in the Lips' collection, is composed of a number of rotating toothed wheels, one set of which is connected to a spindle, projecting from the exterior steel door face and having three hooks, to which the combination rings are affixed. These combination rings, as are shown on the top of the lock, contain five discs, mounted into a round box. Rotation of the toothed wheels in the mechanism is brought about by a cog sector, connected to the spindle. This spindle can be moved in or out by pushing or drawing in order to have each of the five combination discs and the predetermined numbers or letters coincide properly. A marvellously good lock, but the in and out movement of the spindle is a great objection for the proper fitting of the lock to a steel door (Figs. 243 and 244).

After the mechanism has been moved into the locked position the box containing the combination rings can be removed and deposited in the owner's pocket without leaving any indication of the spindle setting.

The letter or cipher combination can be changed ad lib. by means of a small aperture, subsequently protected by a conical disc, which in its turn retains the chosen combination. Many safe door locks of complex design have been marketed, for instance by Pillard, Davidson, Marvin etc., but it would go far beyond the scope of this book to analyse them all.

243 and 244. Isham's register lock. Lips' collection.

CHAPTER VIII
Safe Deposit Box Locks

The use of safe deposit boxes dates from the middle of last century, when leading banks in the U.S.A. offered these boxes, built in their vaults, to the Public for the protection of their valuables. This use of safe deposit boxes gradually spread and nowadays such deposit safes are available in practically all banking establishments in the U.S.A., Europe and elsewhere, except in Great Britain, where valuables are usually entrusted to the banks in "open deposit", and compared with other countries the use of safe deposits there is very limited. In the early stages, ordinary key operated lever locks were used as standard for such boxes, but the increasing demand necessitated a stronger degree of security, which ultimately led to the design of quite specialized locking devices, the construction of which was and has been confined ever since to experts. After the introduction of the safe deposit box system — by the end of the 19th century — European lock manufacturers were wide awake to the popular trend, with the result that they brought onto the market a choice of high grade security locks of the changeable key, interchangeable key, and the key and combination types. Credit must be paid to these European manufacturers for keeping pace with the developments by producing the latest perfection in safe deposit box locks. The writer feels justified in saying this, since he has specialized in this kind of lock for more than 50 years and many patents have been granted to him for locks for this particular purpose. Lips' safe deposit boxes have reached the four corners of the globe. Locks for safe deposit boxes can be classified as follows: a) key operated locks, b) key and combination locks combined, c) keyless combination locks. The types a and b are used mainly in European countries and type c principally in the U.S.A.

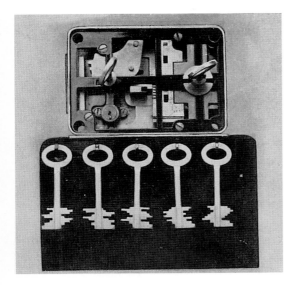

245. Lips' First changeable lever safe deposit box lock in unlocked position.

246. Lips' First changeable lever safe deposit box lock in unlocked position.

Key Operated Locks for Safe Deposit Boxes

Four types of locks fall into this classification:

1. The type that is fixed to the door by means of screws.
2. The type with changeable levers, set by the key.
3. The interchangeable type which can be fixed to any door without the use of screws.
4. The key and combination type of lock combined to form one unit.

Because of the inadequate security supplied by the key lock mentioned under 1, this is no longer used; the new renter of any box so secured was always worried in case his predecessor might have made a duplicate of his key.

The Key Lock with the Changeable Levers

The Lips' changeable lever lock, as shown in Figs. 245 and 246 contains a number of levers of the non springloaded type, which levers are adjusted by a key. Once the levers have thus been set by a key, the lock mechanism can be operated by that particular key only. Locks based on this principle have been made and put on the market by several manufacturers such as Kastner, Sommermeyer, Chubb, Diebold, Sargent and Greenleaf, Yale, Lips, and many others, but they differed in construction as some of these locks contained springloaded levers and others did not.

This changeable lever lock is of a wonderful and ingenious design, such that the writer feels obliged to give a detailed description of its function and operation. The Lips' Patent Springless Lever Lock Type 513, will be analysed first, followed by those of some other makes.

Fig. 245 shows the Lips' changeable lever lock in open position, with both keys inserted in their respective keyholes (the custodian's key on the left, the renter's key on the right).

Fig. 246 shows the lock with the bolt thrown in locked position. For a proper understanding, it should be observed that this type of lock consists of two or more mechanisms in one lock case, one mechanism under the control of the custodian's key and the other under the control of the renter's key. Dual control is thus provided, for unlocking can only take place in the presence of the renter and the guardian. The former will first have to release the mechanism of his lock part, after which the latter with the custodian's key can withdraw the lock bolt and open the door. It is then the custodian's task to lock the box again afterwards. Now suppose this type of lock changes renter, a new setting of the levers to suit the new renter's key can be brought about only by the banker's key, a small sized cylinder key, which is inserted vertically into the keyhole below the custodian's to engage a cylinder behind the lock mechanism near the back plate of the lock case. This cylinder or banker's key is turned in the cylinder after the lock mechanism has been made to take up the open position. This manipulation will set the mechanism to suit the new renter's key, which remains in the keyhole during the setting operation. The final setting of the levers is obtained by a further turn of the renter's key. Another feature of this most remarkable lock is that the custodian's key can only withdraw or shoot the bolt with the renter's key in place, which means that the renter's key controls the custodian's key, so that this type of lock can only be opened or closed in the presence of both keys, the absence of either making the other inoperative. On closer inspection of the reproductions of this lock, the cylinder attachment for the banker's cylinder key is shown vertically below the custodian's keyhole. The Lips' changeable key lock may be considered a maximum guarantee of security and notwithstanding the fact that its levers are not of the spring loaded type, overlifting of the levers is out of the question, the bit controlling the whole.

The principle of this lock design was inspired by the Newell lock, invented nearly a century ago, although its present size is about 1/10th of the original and moreover the modern changeable lock comprises three locks in one lock case.

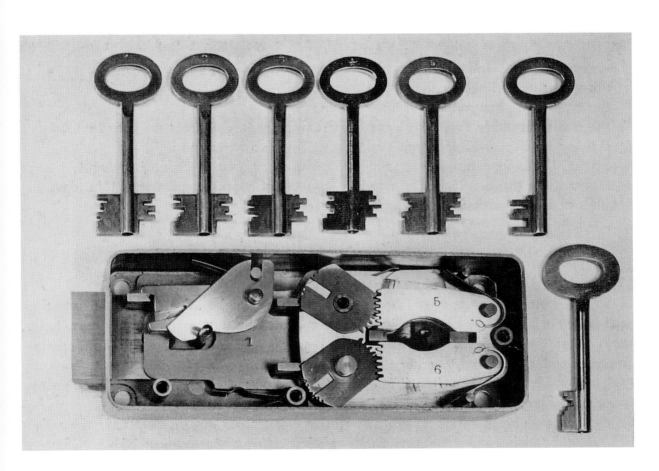

247. Lips' New Changeable key lock for safe deposit boxes (type 3200). Lock in locked position.

248. Lock in open position.

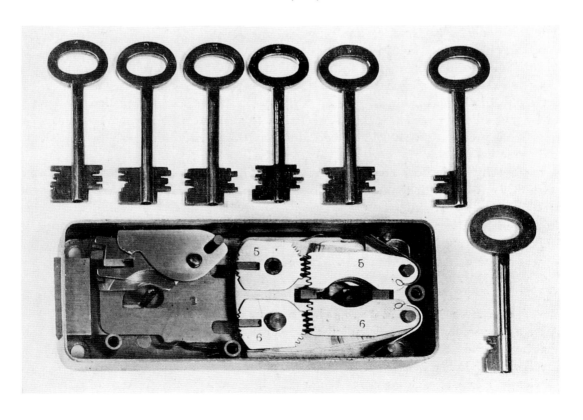

249. Lock in neatral position (Type 3200).

The Lips' Changeable Lever Lock (Type 3200) for Safe Deposit Boxes

The design and construction of this lock, a later achievement of the writer, differ fundamentally from those of the type 513, just described. The mechanism of this lock is shown in Figs. 247, 248 and 249 in its different positions, namely locked, open and neutral. As is usual for this type of lock, it has two keyholes, the one under control of the custodian's key and the other accessible to the renter's key. The mechanism comprises a double set of levers, 11 in total, further, a group of five levers are operated by the custodian single bitted key. However, the renter's key is of the double bitted type and masters the double 11 lever arrangement, that is, the five in the top and the six in the lower section of the mechanism. The number of variations in lock and key arrangements, i.e. the number of permutations, is considerable and easily exceeds 2,000,000 highly practical permutations. Similar to the type 513, unlocking of the renter's part of the mechanism is achieved by turning the renter's key in its appropriate keyhole. Subsequently the custodian keyhole is released and his key inserted and turned to withdraw the bolt. The custodian key is removed with the lock bolt in open position.

A simple turn of the renter's key will not only throw the bolt but will also automatically block the custodian's section of the lock and its keyhole.

Setting this lock mechanism to take up a new permutation is a manipulation characterized by precautionary measures consciously taken to guarantee the maximum security, for it needs the presence of three specific keys at the same time, viz. the banker's, the custodian's and the renter's key.

A new setting of the lock mechanism can be obtained only with the lock in the neutral position, i.e. when with the aid of the banker's key, the renter's key — after opening the lock — has been removed from the keyhole. Any double bitted key arbitrarily selected from the banker's collection, but with the incisions of the key bit differently arranged, will, when inserted into the keyhole and turned, adjust the lever mechanism to suit this key, after the banker's key has been removed from its keyhole. After this new setting the lock will reject the preceding key.

The illustration in Fig. 249 shows the lock in the neutral position; the cams or notches in the edges of levers No. 5 and 6 stand clear of those of disc No. 5 at the left.

After a new setting of the mechanism the notches of the levers and discs will gear, so that by a further turn of the key the gatings are placed in line with the bolt stump, as the illustration in Fig. 248 shows.

Obviously, a lock of such ingenious design must be considered the outcome of a thorough study by experts.

The outstanding features of the lock under discussion may be summarized as follows:

1. Simple operation. Custodian key is operative only after the renter's key has been inserted and turned.
2. The custodian key withdraws the bolt and is removed.
3. A simple turn of the key in the right direction will secure the bolt.

4. The renter's key can be withdrawn only from the lock in closed position, so that there is every guarantee that both sections of the entire mechanism are safely locked with the key in his custody.
5. The setting to a new and differently arranged double bitted key can be effected only in the presence of the banker's, the custodian's and the renter's keys.
6. For a new setting of the lock the renter's key has first to open the lock, the bolt is subsequently withdrawn by the custodian key. It is only after these operations that the banker's key can bring about the new setting. The bolt is fully withdrawn by turning the banker's key, thus releasing the discs, after which operation the renter's key can be removed from the keyhole with the lock in open position. By inserting and turning the new renter's key the entire lock mechanism will automatically adjust itself to this new key and reject its predecessor.
7. Precautions have been taken to prevent the key from being inserted in the wrong way, which might dislocate the mechanism.
8. A new renter has every guarantee that his predecessor does not possess a duplicate of his key.

To the writer's knowledge there is no safe deposit lock that can claim identity to the Lips' safe deposit box lock.

The Interchangeable Key Operated Lock

Under this heading the Lips' lock type 503 will be dealt with as representative of this class of lock.

Changeability of a lock means the setting of a lock to suit a new key neutralizing the use of the preceding key (Fig. 250 and 251).

Interchangeability of a lock means that the complete lock can be removed from the door and replaced by another lock operated by an entirely different key. For this change no screwing, boring or cutting is required, as this operation is done simply with the aid of the renter's key. Consequently the security features equal those of the changeable lock type, for in both cases the renter of the box holds a different key. In what manner is this change of lock accomplished? The renter's key is inserted into the keyhole and then turned till the key bow has reached the vertical position. By doing so, a slide fixed to the back plate of the lock case is released, which permits the entire lock to be removed from the steel door of the safe deposit box and to be replaced by another lock, which is fixed to the inside face of the door, sliding on four pins as guides. Subsequently the slide is pushed into the locked position and the lock case is now firmly and securely fixed to the door. The new renter has every guarantee that the lock is under his exclusive control and that copying of his old key by his predecessor is made ineffective.

The Lips' interchangeable lock, type 503, has been widely used for many years and hundreds of thousands are in use in all parts of the world.

The renter's lock section comprises a double set of levers actuated by a double bitted key and the number of key variations or permutations runs into millions. This type of lock is available with either springloaded or springless levers, according to the requirements and wishes of the Bank. A springless lever lock of the type 3101 is shown in Fig. 252 that equals, if not surpasses, the springloaded lever lock in security, although the springless lever type of lock does not give the user that pleasant sensation of smoothness when turning his key, which a springloaded lever lock does. The springless lever lock contains a set of upper and a set of lower levers. These two sets of levers are interconnected by the tail ends, marked A, and the resultant co-ordination of lever movements assigns the proper positions to these levers and prevents them from overshooting their proper positions by too quick a turn of the key. This security feature is characteristic of all the Lips' safe deposit box locks.

The Lips' interchangeable safe deposit box lock, type 3103, although smaller in size, actually offers the same high degree of security as that of the type 3101, being especially designed to meet the growing demand for smaller sized safe deposit boxes.

250. Front view of a safe deposit box door. Lips' collection.

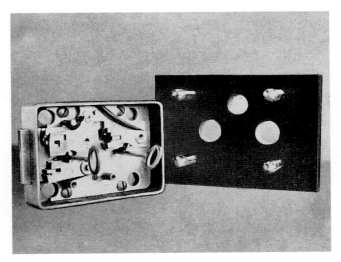

251. Inside view of a Lips' safe deposit box lock with front plate removed.

Key and Combination Locks Combined

The highly complex design and construction, peculiar to this class of lock, demands a detailed analysis, and since the Lips' lock type 501 can be considered an outstanding example of its class, it will be used for this purpose.

To combine the properties of a key operated and a combination lock and to accomplish their mechanical realization, so as to constitute one compact unit in a single lock case, is the result of many years' research and study. A glance at the reproduction in Fig. 253 of the Lips' key and combination lock, type 501, will convey to the reader an impression of its complexity.

Each of the 84 moving parts (either sliding or rotating and all machine made) performs its own task in the mechanism, with the result that perfect security is attained. Besides the two key operated lock parts, the mechanism comprises a four dial combination lock. This lock type can be more easily operated by safe deposit box clients as a single dial lock.

The renter's section of the lock contains a double set of levers, operated by a double bitted key, whereas the custodian's key is of the single bitted type. The renter's key can only be inserted into the keyhole after the custodian has opened his section of the lock and has thus made the renter's keyhole accessible. However, the renter's key cannot be turned until the letters on the four dials have been brought opposite the setting mark in their predetermined order. In designing this lock, the writer was guided by the principle that an open door showing the proper letter combination, which led to its unlocking, should be avoided by all means, as this would enable another renter to read the combination. Therefore, the renter's key must first make a quarter turn, then the fixed letter combination is thrown off or dislocated and it is only after this manipulation that the renter is enabled to turn his key fully to open the box door. When locking the box with the aid of his key, the renter need not worry about the position of the four dials, as the combination lock closes automatically and simultaneously with a turn of the renter's key in the locking direction. The little knob in the centre of the combination lock serves to neutralize the guardian's key and to block the renter's keyhole to make it inaccessible to any other key. The letter combination can be changed at will in a very simple way without the lock having to be dismantled or removed from the door. Such a change in the combination can be done by a layman, no expert knowledge being required, as the lock can be considered fool proof in this respect; even if unexpectedly a mistake should be made disturbing the mechanism, this can be rectified by simply pushing a special pin on the back plate of the lock. The numerous variations of locks based on the same principles excludes discussing them all. Yale and many other American lock manufacturers produce cylinder locks for safe deposit boxes, which, however, are not in very great demand in Europe, neither are the changeable safe deposit box locks, shown in Fig. 254 and 255 operated by single bearded flat cylinder keys.

252. Lips' springless safe deposit box lock. Lips' collection.

253. Lips' key and combination safe deposit box lock. Lips' collection.

254. Yale's safe deposit cylinder lock. Lips' collection.

255. Yale's changeable safe deposit box lock and its corresponding flat keys. Lips' collection.

256. Rosengren's safe deposit box lock. Danish. Lips' collection.

 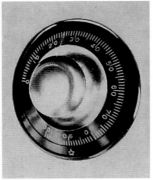 

257. Dismantled, showing the four combination discs, each sub divided into a scale of a hundred numbers.

258. The outside dial and knob.

259. Assembled.

Europe is undeniably biassed towards the double bitted renter's key and its inherent innumerable variations or permutations.

Sargent and Greenleaf applied flat white metal keys to their screw fixed locks. The firm of Rosengren of Goteborg, Denmark, manufactures safe deposit box locks of the interchangeable cylinder type, but their use is practically confined to the Scandinavian countries only.

An illustration of the Rosengren lock is found in Fig. 256. The change of cylinders in this lock also requires the presence of the custodian's and the renter's keys simultaneously. As already mentioned, the system of safe deposit boxes has so far been introduced on a very limited scale in England and is actually restricted to some vaults in London Banks. Consequently Great Britain may be considered an important outlet for such installations. The safe deposit box locks actually in use are, for the greater part, imported from the European continent.

Germany, Holland, Belgium, a.o. European countries, have been very active in the manufacture of safe deposit box locks of the interchangeable type and many important banking institutions have special safe deposit vaults at the disposal of their clientèle.

With regard to France the same types of safe deposit box locks are found there as were in use more than 50 years ago.

In conclusion, some special features of the Yale and Towne safe deposit box lock should be mentioned.

The levers of the key operated changeable lock are divided into two separate parts which, when geared, swing on one common pivot, which in its turn is under the control of the banker's setting key. The pivot functions as a kind of eccentric to separate the two lever parts, of which the edges are provided with notches or cams.

On closer inspection of the illustration of the Yale and Towne lock the lever arrangement as described can be seen. The position of the lever bellies in this lock is changeable to match the gatings through which the bolt stump will slide when turning the key.

The Sargent & Greenleaf Company and several other firms manufacture locks based on this principle (Figs. 257, 258 and 259).

CHAPTER IX
Some General Remarks on Keyless Combination Locks

In the July 1925 number of the *"Scientific American"* the headline of an article by Mr. E. J. Goodnough ran as follows:

"There are no 'Jimmy Valentines', the burglar does not live whose sense of touch or hearing can find the combination of the modern vault".

In this article Mr. Goodnough, an expert of one of the leading American lock factories, reports his experiences regarding the tales circulating throughout the U.S. and, for that matter, all over the world, that high quality combination locks could be opened by the sense of touch or hearing. He writes:
"Some months ago I was on my way to a town in Tennessee to visit a bank that had reported the combination lock of its vault had failed. One of my fellow travellers spontaneously drew my attention to an article in his morning paper, exclaiming: 'Of all the bunk! This article tells about a safe having been blown and the guys getting away with $12,000. What I can't figure out is why the fellows who do this sort of thing, don't use their heads a little more'.
He then went on to say: 'If they would spend their time studying combination locks they would soon learn to open them by sense of touch or hearing without disturbing the combination of the doors and running the risk of injury or arrest because of the racket'.
It was evident from the description in the paper that the combination lock of the vault was one of outstanding quality and solid ingenious construction. I told my friend I had often heard about smart fellows who could open any combination lock in this way, but that I had never had the privilege of meeting such a crack, as otherwise I should have asked him how he managed this.
'Well', my companion replied, 'it takes a lot of time, study and patience in particular. In nine cases out of ten I can do it myself. It's all the result of cultivating a keen sense of touch, so as to feel the slots in the tumblers as they come in contact with the unlocking device. In other cases you can hear the tumblers as they fall'.
I was all ears and after he had related several of his experiences and technique in opening safe and vault doors of the burglar proof type, I asked him what his business was. 'Oh, I'm with a burglar alarm company', he replied. I asked him to mention some names of banks, the vault of which he had opened in this way. Most of them were known to me, and one in particular as it had a ten inch thick solid steel door and was equipped with the latest time and burglar proof combination locks. I told him I was afraid it was all nonsense he was telling me. 'I have spent all my life in the safe business', I said, 'and I know it is absolutely impossible to do what you are claiming. It so happens I'm on my way to a bank where the combination lock was reported to have stuck. Now if you are willing to accompany me, I will pay you $1,000 if you open the locks (under the most favourable conditions) without disturbing the mechanism of the door'.
He hesitated a little, but finally accepted saying:
'I guess you've got me. I had talked about this sort of thing so much in my business that I've begun to believe it myself'.

And so it is with authors of detective stories, who in their own interests promote certain lines of business by thus hammering away at the integrity of the highest grade combination locks. For this reason, it is not hard to explain why the public at large and even some bankers have an exaggerated idea of the capabilities or skill of crooks. Those who are telling such tales are merely repeating hearsay, or rumours, trying to expound their knowledge even though they are merely laymen in this field. Unfortunately, it cannot be denied that cheap lock types on some fire proof safes can be opened by the sense of touch, reading and hearing.

Cipher combination locks and real key operated safe locks, especially those of the double bitted type, supplied by firms fully alive to their responsibilities, successfully defy opening by sense of touch or hearing, under regular conditions.

To-day, some years after writing the first issue of my book, new features and new constructions in many and varied fields have been produced. It is impossible to mention all of these, but in the field of combination cipher locks noteworthy and ingenious new patents have been filed.

All experts are aware that combination cipher locks with four tumblers like those made up to 1950 offered great safety, manufacturers exerting their full attention to this.

A few experts succeeded in finding a method based on reading and hearing to develop cipher locks with three tumblers. The requirements for these were, however, proportional to their safety, and in the opinion of many, a burglar has never the favourable circumstances of being able to work quite quietly, to make all kinds of notes, to work without intrusion and where absolute stillness reigns. It is therefore not uncommon for the burglar in constant fear of detection to find that long before trying a fraction of the possible combinations his time has expired.

We have so far referred to cipher combination locks with three tumblers, but with four tumblers the position becomes still more difficult and the chances of success much smaller. A combination lock with three tumblers makes about 1,000,000 combinations possible, whilst such a lock with four tumblers raises the possibility to 100,000,000.

The degree of finish of the lock's component parts is also very important, that is whether the component items fit accurately with only minimum tolerances for working or whether they are made as a "rattling fit" due to a widening of the manufacturing tolerances to cheapen production.

There are several recognized experts in this field in the U.S.A., like Robert Murray, New York, N.Y., and Harry Miller from Rochester, New York, who during the last few years have made public their opinions, and also those who have offered solutions to prevent the opening of locks by unauthorized persons.

The writer has also applied for patents on a new construction, which was very favourably judged by experts in the U.S.A. during various lectures on this subject. Whilst we are hoping to obtain good results with our invention at some future date, it is in my opinion too early to further enlarge upon it here.

Letter Combination Locks

The construction of the early letter combination locks, manufactured in several European countries, comprised 3 or 4 knobs or dials.

In the days when the manufacture of safes was not the specialized art it is now, a number of blacksmiths produced safes with letter combination locks of their own design and construction. Each of the knobs or dials had its own spindle, connected to the inside combination disc, and sub divided into 24 sectors. In addition this disc was controlled by a tooth wheel connected to a spring loaded tumbler, to prevent it from shifting to another combination. A new setting of the letter combination was obtained by raising the tumbler. This type of lock, the principles of which have been discussed in the preceding pages, has become obsolete due to the risk that any unauthorized person could read and change the letter combination with the door in an unlocked position. Besides, the four knobs on four spindles when torn from or punched through the door by force, would leave an equal number of holes right through the entire thickness of the door. The same thing would happen if the knobs were removed from the spindles with a screw driver, so that the spindles could be hammered through the door.

There was rivalry among American lockmakers to improve the construction of this type of lock, with the result that in the middle of last century the American industry succeeded in constructing a combination lock with only one spindle to operate 3 or 4 dials simultaneously. It was then that manufacturers like Yale, Sargent, Good, Evans, Watson, Hall, Diebold, Pillard, Jones and others competed heavily in designing highly ingenious but complicated cipher combination locks, recommended as the optimum of security. Some specimens of these cipher locks will be described and reproduced on the following pages. If I remember rightly, it was Theodore Kromer of Freiburg who constructed a one-spindle letter combination lock. He was a watchmaker and had been in the employ of several lock manufacturers in the U.S.A., all concentrating on the cipher combination lock, but after his return to Europe he concluded that the Continent was biased towards letter combination locks. Even though Kromer's letter lock was of a highly ingenious construction the Sargent cipher lock showed important advantages over it, also a greater number of combinations were permissible in the U.S.A. locks. Even so, it is worthwhile inspecting the basic principles of Kromer's lock more closely, since there is no doubt about its being one of the better class letter locks.

Kromer's Letter Combination Lock

The lock case contained three combination discs or tumblers, each provided with 50 indentations. Two of the knob operated dials had an alphabet of capital letters and of small letters, so that the number of combinations possibly amounted to $50 \times 50 \times 50 =$ a total of 125,000, such being considerably lower than the 100,000,000 combinations of a modern four tumbler cipher lock. Despite this, Kromer's lock may be considered one of the best letter locks in Europe, even though it has gradually given way to the cipher locks. Kromer's lock is shown in Fig. 260, whilst Fig. 261 shows the lock with the back plate removed to expose the mechanism.

Over the combination tumblers a lever is shown, balanced on a centre pivot through an oval centre piece. This lever rested on the edges of the combination tumblers and was lowered when the oval part engaged the notches in the edges of these tumblers, when turned, or were raised in places where such notches were absent in the tumblers. One of the side faces of the lock case shows a round hole near the top end to receive the locking pin, attached to the locking mechanism of the safe door. This pin could only be inserted when the tumbler was in the lowered position, whereas this hole was blocked when the tumbler was raised. Since the lever was balanced by the central pivot, the combination tumblers could be turned to take up the locked position before the safe door was actually closed. As soon as the locking pin was removed the spring loaded tumbler would automatically lock. Experts have their doubts about the efficiency of the balanced tumbler, as a powerful bump or shock in vertical direction might throw the tumbler into the unlocked position.

The spring of the spring-loaded outer dial was as provided with a pin which, when the lock was set to the proper combination, caught one of the 50 indentations in the dial on the inside of the lock case, so that the inside and outside dials were interconnected and formed one unit. Both dials were operated by a letter knob, fixed to the outer face of the door, which knob was connected by a steel spindle to the discs on the inside of the lock case. Incorporated in the mechanism close to the point where the spindle and the tumblers coincided, were spring-loaded bevelled pins, engaging the notches in the tumblers. To operate the combination, supposing it was set to read H.O.L., the dial had to be pushed in and given a full turn to the right as far as H, then pulled out and given a further turn till the letter O was reached, and finally, turned to the left to the letter L. These operations being completed, the lever is dropped, the mechanism released and the lock opened.

260-261. Kromer's letter lock. Lips' collection.

The combination was changed by first setting the combination actually in use, and a round pin was subsequently pushed into an opening in the back plate of the lock case, thus releasing the tumblers to allow them to be set to a new combination.

The letter lock, it cannot be denied, had excellent mechanical features, nevertheless it was entirely superseded by the cipher combination lock. Some years later Kromer acknowledged the superiority of the cipher lock by bringing on to the market cipher combination locks, inspired by and designed on the principles of the American cipher lock system. The use of cipher combination locks gradually extended and to-day this type of lock may be considered to have a monopoly of the European market. Only the southern countries in Europe still keep to their old construction.

Cipher Combination Locks

James Sargent of Rochester, N.Y. was one of the best known and leading lock experts of his time. His craftsmanship was born out not only by his great skill in picking key operated locks, but also by his reputation as an expert in opening combination locks. For this "work" he used a micrometer, an instrument of extreme sensitivity, designed and constructed by himself in 1857. This instrument is in the John M. Mossman Collection of the Museum of the Society of Mechanics and Tradesmen in New York. Sargent's micrometer, shown in Fig. 263 (together with the first Sargent lock) could measure the ten thousandth part of an inch with great accuracy and has proved to be a useful and efficient instrument for picking combination and key locks of any description and he once won $1,000 by opening a combination lock. His keen knowledge of the vulnerable spots in the locking devices actually in use, stimulated him to design a lock which was

262. Sargent's original cipher combination lock.

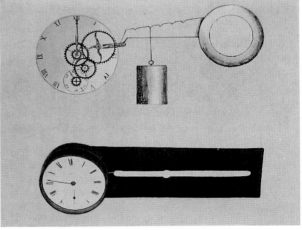

263. Sketch of the basic design of Sargent's micrometer.

proof against any opening efforts, even those by means of a micrometer, and extensive experiments resulted in the construction of the first cipher combination lock to resist an orthodox opening. The first Sargent Lock is ilustrated in Fig. 262. However, the news of Sargent's achievement in 1865 spread like wildfire and his competitors, also constantly on the alert and anxious to bring out improved combination locks, soon followed suit. For a further description of the construction and operating principles of these locks, the writer has made a selection from the older as well as the modern types of this class of locks.

It may be said without any exaggeration that even today Sargent's original combination lock is considered one of the best security locks ever made and still made by Messrs. Sargent & Greenleaf Company of Rochester, N.Y.

The construction and operation of all cipher combination locks are based on the setting of combination tumblers to a certain predetermined cipher combination, any other position of the tumblers preventing the lock from being opened.

Pillard's Cipher Combination Lock

This product, invented by Mr. J. Pillard of New York in 1868, was manufactured by the New Britain Bank Lock Company of New Britain (Conn.) The lock contained four tumblers, the combination of which could be changed arbitrarily by a special setting key. A heavy lever fitted over the discs, and served to block the bolt provided the proper combination was thrown out.

Yale's Double Dial Pull Combination Lock

This magnificent lock, reproduced in Fig. 264 is in the writer's collection and is due to the ingenuity of Mr. Linus Yale (1867). The lock comprises a double set of rotating tumblers, each connected to a separate dial by a spindle. Both dials serve to unlock and to lock the mechanism. The second dial was also used in emergency cases should somebody find himself locked inside the vault. The illustration of the lock mechanism shows two cylindrical boxes, both containing four tumblers. The boxes form an integral part of the lock case and cannot be detached or adjusted. The covers of the boxes can only be taken off with the aid of a special key—the banker's key—so that a re-setting of the tumblers can only be effected under the banker's supervision.

Dexter's Double Dial Combination Lock

A product of Messrs. Herring & Co., lock manufacturers of New York. This lock, dating from 1869, contained a double set of combination tumblers like Yale's and was similarly provided with two dials. Each of the two bolts was moved by its separate set of tumblers. This lock is a masterpiece of ingenuity and precision and later on Dexter added a time lock to the mechanism, so that the combination could not be worked until the predetermined period had elapsed. Lock in Fig. 265 from the writer's collection.

264. Yale's double dial combination lock. Lips' collection.

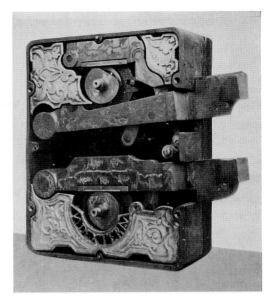

265. Dexter's double dial combination lock. Lips' collection.

Jones' Patent Combination Lock

This remarkable lock was designed by H. C. Jones of Newark, New Jersey, in the year 1849 and differs from other American cipher locks of this class in the number of tumblers. It had four dials, each bearing ten numbers which could be brought into line with six different setting marks, whereas the knob in the centre had to be brought into line with only one of the six setting marks before the other knobs could be turned. This arrangement meant a considerable increase in the number of combinations. After setting the four dials the knob in the centre was brought into alignment and pushed in, which manipulation releases the bolt. Lock in Fig. 266 from the writer's collection. Although a lock of ingenious design and construction it has fallen into disuse. Modern cipher locks are all of the mono-knob type, and after the combination is set the bolt can be moved in or out.

Dexter's Single Dial Combination Lock

The operating principles of this lock are similar to those of Dexter's Double Dial Lock. The conical spindle ground into the thickness of the steel door of the vault was a precaution against it being torn from or hammered through the door. This lock had four tumblers which could be re-set to form new combinations. (Fig. 267 and 269).

Lillie's Patent Combination Lock

In August 1865 Lewis Lillie of New York was granted a patent for his new combination lock. Like Kromer's lock it contained a spindle connected to a dial, which had to be pushed in or pulled out to work the combination and move the mechanism. The combination was set by pushing in the knob and by turning it to the first number of the combination, then the knob was pulled out and turned to release the lock bolt. The advantage of this system was that the combination was thrown out the moment the locking and unlocking was brought about. Lock in Fig. 269 from the writer's collection.

266. Jones' patent combination lock 1849. Lips' collection.

267. Dexter's single dial combination lock. Lips' collection.

268. Lillie's patent combination lock. Lips collection.

269. Dexter's single dial combination lock. Lips' collection.

The Lips' Cipher Combination Lock

When designing this lock the Lips' factories were guided by the principles of the Sargent Combination Lock. For use in fire proof safe doors it contains three combination tumblers, although four tumblers are used when this lock is to protect heavy armoured safe and vault doors, which have indirect spindle.

Each disc bears a scale of 100 changeable numbers so that when the three tumblers are used the number of combinations amounts to 100 times 100 times 100 = 1,000,000 and in the case of four tumblers this number runs up to 100 times 100 times 100 times 100 — 100 million combinations. As is evident from the drawings in Fig. 257 the tumblers are composed of an inside and an outside disc, the one revolving inside the other. A special feature of the Lips' Combination lock, and incidentally not found in any other lock of this classification, is that with only one quarter turn of the dial in the locking direction, all combinations are automatically thrown off. Unlike the Lips' lock, dials of any other type of lock have to be revolved at least four times to achieve the same effect.

The tumblers are not made of cast, but of pressed bronze, a die process that guarantees absolute equilibrium of the tumblers and avoids oxidation. The manufacture of these discs is a real precision job.

The edge of the inside disc (a) in Fig. 270, is provided with minute cams, whilst the outside disc has two sets of engaging teeth, the tail ends of which are supported by an eccentric device. The edges of the top ends of these teeth have a range of cams which mesh with the indentations of the inside disc when the eccentric device is revolved. By this manipulation the inside and outside discs are interconnected, i.e. they are reduced to a single tumbler. The outer disc shows a slot (e) into which the fence of the lock bolt drops after the disc has been properly set to open the lock. Each inner disc has a square peg, interlocking the adjacent disc. So in the case of a lock containing four tumblers only the front one will revolve at the first full turn of the dial, the second full turn of the dial will take along the second tumbler, so that after four full turns of the dial the four tumblers will revolve synchronously. A notch screwed to the spindle of the dial conveys the rotary motion of the dial to the tumblers. Each disc contains a flyer, so that loss of combination numbers on account of the size of the square pegs is avoided and the use of the full range of numbers from 1 to 99 is guaranteed.

The flyer on the spindle has been shaped to prevent the lever from settling on the combination tumblers. Only when all the four tumblers are accurately set in the unlocking position will the lever or fence catch, so that a lock of this description may be regarded as very secure.

As already stated, the tumblers are of bronze, while the lock case is sometimes made of steel for fireproof safes and of cast bronze for all other purposes, bank locks etc. Fixed non rotating rings separate one tumbler from another, thus avoiding points of contact between the tumblers, and possible disturbance of one tumbler by the motion of setting another.

The spindle, made of high grade and often heat treated steel, is divided into stepped sections (see Fig. 273) and is incorporated into the thickness of the door from the inside and secured by a filling-in piece. The spindle may also consist of conical sections, which type resists tearing, punching or hammering. In addition the spindle is often provided with steel pins, arranged criss-cross, to prevent their being drilled out. Lips' spindle constructions of this nature may vary in design to suit particular models and sizes of safes and vault doors (Fig. 271-274).

Finally, the locks are of the direct or indirect spindle type; in the former the movement of the spindle is imparted to the tumblers without any medium, whilst in the latter a transmitting cam wheel is inserted with the same effect. This latter type with a four-tumbler combination lock provides maximum security.

Since the great majority of cipher combination locks have a universal system of operation, the system described here can be considered applicable to that class of lock in general. Supposing the combination: 25 - 75 - 38 - 94 has been chosen for a right hand lock containing four tumblers in locked position, at least five full turns to the right will be required, and if the initial number is ignored, four times. After the fifth turn the number 25 is brought dead opposite the setting mark. The question arises: why five full turns? In describing the lock mechanism the special cam was mentioned which, screwed to the spindle, serves to convey the rotary motion of the dial to the tumblers.

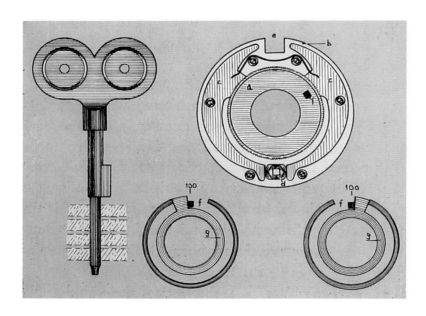

270. Inside combination disc.

Left: Banker's setting key.

Right: Component parts of the Lips' combination Disc.

The position of the tumblers after the lock bolt has been thrown is entirely neutral and arbitrary with regard to the setting of the combination to unlock the mechanism. Consequently after five full turns of the dial it is certain that the cam will have seized the first tumbler to be set, will subsequently seize the second, the third and the fourth. So, after five full turns to the right, the number 25 is brought in line with the setting mark, after the next 4 consecutive turns to the left, the number 75 is placed in alignment, one full turn for the cam to seize the first tumbler and three turns to bring the second one in the unlocking position, followed by three turns to the right to engage the number 38, finally to stop at 94 after two turns to the left.

Now the setting of the combination has been completed, i.e. the slots of the tumblers are in line, so that the lever drops into the row of openings in the tumblers. By one final turn to the right the lock bolt can be withdrawn.

The following schematic summary may be considered the unlocking formula:

5 turns	4 turns	3 turns	2 turns	1 turn
right	left	right	left	right
25	75	38	94	open

For a left hand lock the order of turnings to the right or left is reversed, consequently five initial turns to the left etc., etc.

To change the combination it will be necessary to set the discs on the combination actually in use, in the manner just described. However, the fifth manipulation, resulting in the actual opening of the lock, is dropped. Only with the lock in this position can the banker's key be inserted into a keyhole at the back of the lock case to bring about the new combination.

When locking the safe or vault door, the dial is turned five full times to the right, so as to make sure that the combination has been thrown out.

Later on the Lips' Cipher Combination Lock was perfected and patented by including a device which automatically threw out the combination by a quarter turn of the dial only and which added to the convenience of locking and unlocking.

This feature, for which a patent was granted to the writer, covering the Netherlands, Gr. Britain, France, Belgium. U.S.A.. Sweden and other countries, has greatly enhanced the security of this lock.

271. Lips' New Combination lock. "Patent applied for."

272. Inside view of the Lips' Cipher Combination lock.

273. Spindle with stepped sections incorporated into armoured vault door.

274. A set of combination discs.

Chronometer Time Locks

It was in 1872 that the manager of a bank in Great Barrington (Mass.) U.S.A. was aroused from his bed by burglars and forced to go with them to his bank. There he was told he could either disclose the combination to operate the lock of his safe or be shot. Such "hold ups" frequently occurred and lockmakers felt called upon to counteract these methods by making still better locking devices. These efforts resulted in the invention of the time lock. Mr. James Sargent must be given full credit for this invention, which turned out to be a successful and conclusive remedy. Sargent's patented time lock was first practically tested when it was mounted on the vault door of the First National Bank of Morrison (Illinois) in 1874 by Mr. Sargent himself. The price of that time lock was $500.00, but it was worth many times that sum.

The time lock is essentially an unlocking device, used in conjunction with the combination lock and is meant to prevent any one, including the bank manager, whether he knows the combination or not, from opening the door until the hour for which it was set arrives. The illustration of Sargent's first time lock in Fig. 275 shows the mechanism in a strong bronze case. It comprises two chronometers and a rotating disc with a gating or slot in its edge. For the operation of unlocking one chronometer will do, since the other serves as spare, should one refuse to function.

Both chronometers could be set to run for 2 x 24 hours, a maximum fixed, to cover the eventuality of closing down for two consecutive days (a bank holiday followed by a Sunday or vice versa). The operation principles are very simple: Supposing a banker locks his vault door at 6 p.m. till 9 a.m., a lapse in time of 15 hours. In this case the chronometers are set to run for 15 hours.

The bolts are thrown by means of the heavy handle on the door front. This done, the time lock is automatically locked, the disc with the gating or slot is displaced, so that this gating cannot receive the locking bar and the levers of the bolt system cannot drop into the row of openings.

Since Sargent's invention, time locks have undergone many changes and improvements, thanks to the ingenuity of many inventors in this field, whose products have proved highly efficient in practice. Today timelocks are made with three and four chronometers, capable of running over 120 hours.

275. Sargent's first Chronometer Time lock (1872).

Some of the first time locks, dating from the latter part of the 19th century, will he illustrated and described, followed by some interesting details of modern time locks and automatic locking devices from the writer's collection.

Though different in design and construction, and each showing particular features, all time locks have one principle in common, namely the absolute control of the opening hour. Statistics have disclosed that the number of time locks in use throughout the world exceeds 75,000.

Pillard's Time Lock

This lock was designed and constructed by the New Britain Bank Lock Company in 1877 and contained two chronometers, each of which could bring about the unlocking, one serving as a reserve in an emergency.

The lock comprises two locking bolts, one working in a vertical, the other sliding in a horizontal direction; in the locked position one bolt is blocked by the other. The chronometer withdraws the horizontal bolt, so that the vertical lock bolt, obstructing the gating of the locking bar, can drop to release the gating. A night and a day setting of the lock can be arranged with the aid of a knob on the inside of the lock case. For intervals longer than 24 hours a second turn of the knob will add the number of hours to make up the lapse of time required, which means that after the first 24 hours the chronometers will bring about the raising of the lever, however, without the slightest effect on the locked position of the horizontal bolt. A specimen of the lock is in the writer's collection (Fig. 276).

276. Pillard's Time lock (1877) Lips' collection.

The New Holmes Electric Time Lock

The Holmes' lock, patented in April 1872, and manufactured by Messrs. E. Howard & Co. of New York, is likewise operated by two chronometers fed by an electric low voltage battery, which in case of emergency, should both timers refuse, can be unlocked by means of a special device. The lock in Fig. 277 is in the writer's collection.

This lock could be set for the day and the night by means of a dial. Supplementary patents on the improved construction of this lock were taken out in 1879 and 1880.

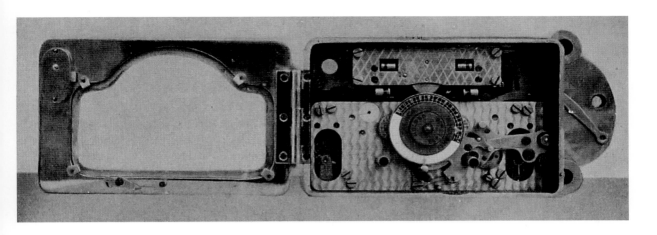

277. The new Holmes Time lock (1872). Lips' collection.

The Dalton Dual Time Lock

A highly ingenious invention, which has found widespread application, not only controls the mechanism of the main lock, but also the cipher combination locks.
Dual control systems have been elaborated by Beard, Yale, Young, Brass, Huston, Hollbrook and many other inventors, but all these systems were based on the same principle, namely the absolute control of the unlocking operation by means of chronometers. In Wooley's dual time lock a minutely regulated flow of liquid forced a counter weight to fall down by force of gravity, following which the combination could be worked.
The Lips' factories have been suppliers of time locks for many years and a considerable number have found their way into safe and vault doors, both of the square and circular types. Yet the application of time locks in Europe which is still limited at this stage does not justify the manufacture of special chronometers. Another point to be considered is that the Lips' factories, in view of their reputation, insist on supplying the very best in this field and consequently have made agreements with a firm of specialists for the supply of time locks (See Figs. 278 and 279).
These two reproductions show the Lips' time lock, in the unlocked position, so that the locking bar can slide into the bolt mechanism and in the locked position, so that this bar is arrested by the locking tongue.
The three chronometers, clearly visible in the lock cases, can be set to run three, four or even five times 24 hours. The fact remains that Sargent and Yale, pioneers in the field of time locks, are considered even nowadays the leading experts in the U.S.A. and their respective companies have been the principal suppliers of time locks for years, so that by far the majority of bank vault doors in the U.S.A. have been fitted with locks of their manufacture.
Time locks are supplied with two, three or four chronometers, according to the requirements of the bank. The time lock of the three chronometer type is in demand to the greatest extent, and is used both for the main and emergency doors of the vault, so that a six-fold security for proper functioning of the vault is obtained.

278-279. The Lips' Time locks (Sargent system).

Automatic Locking Devices and Time Locks Combined

In the present advanced stage of lock technique a first requisite is to prevent holes right through the entire thickness of armoured vault doors. Such holes, either keyholes or holes made to receive spindles of combination locks, have proved to be vulnerable spots and burglars have broken into vaults by attacking these vulnerable spots by methods known as punching, drilling and even burning and blowing.

The present demand is for absolutely solid doors for large vaults, provided with a strong automatic locking device, built into solid thickness of the door from the inside and controlled by a time lock.

The automatic device interlocked with the time lock, as the name already conveys, is entirely self-acting, and its motive power will move the bolt mechanism into the locked and the unlocked position at hours for which the time lock is set.

The operation principles of the automatic device are shown in Fig. 280. On the left is the mechanism of the automatic device in the locked position, whilst the view on the right represents the automatic arrangement interlocked with the time lock in the unlocked position (Fig. 281). The heavy lock case is fitted to the inside face of the door by means of strong steel bolts, whilst the sliding locking bar is connected with the bolt mechanism.

The time lock has been mounted resiliently on top of the lock case, so that shocks to the lock mechanism when moved do not impair the chronometers.

The automatic device is wound by means of a lever, fitting the square pin (A), with the door in open position and after the time lock has been set to run the predetermined number of hours. The automatic device has a horizontal bar, provided with a tumbler at one end, projecting from the edge of the vault door. The moment the edge of the door joins the jamb, this tumbler is pushed in by the door jamb, so releasing the lock mechanism for the final shooting of the bolts.

However, there is a lapse of about two minutes between the release of the mechanism and the final shooting of the bolts, because this interval is needed to allow the fully wound device to run down before the bolts can be fully thrown. This interval of time has been adopted for very special reasons, for it will enable the bank employees to operate the compressor bar wheel on the outside front of the door and to push the door fully into its tight fitting jamb, no crack being left between the door and the jamb. Synchronized action of the automatic device and the bolt mechanism would cause serious damage to the entire locking gear, for the bolts would be arrested by the door jamb, as there is no time left for the bank employee to push the door fully into the jamb. The door and the jamb fit so tightly that they may be considered hermetically sealed, which will repulse any effort to insert explosives into any possible crack, a method often applied by burglars.

When the opening hour arrives, tumbler (B) of the time lock actuates the catch, projecting from the automatic device, to withdraw the bolts and open the door. The reader can conclude that thanks to the use of such devices and time locks combined, the method of drilling holes through the solid thickness of vault doors to attack locks has been entirely abolished. The entire lock mechanism is mounted on the inside face of the door.

The illustrations in Figs. 282 and 283 show specimens of the smaller type of vault doors, the one picture representing the lock in the open position, the other one showing the same device in the locked position.

280. *Exposed view of the Automatic Device in locked position.*

281. *Exposed view of the Automatic Device in unlocked position.*

These reproductions clearly demonstrate, too, the way in which the main bolt mechanism and the automatic device have been interlocked.

Vault doors may be square, rectangular or circular. Tight fitting of square and rectangular shaped doors is accomplished by lining the indented jambs with a bronze packing, whereas round doors are made to fit their jambs hermetically by a high precision grinding process.

The foregoing description of the automatic device will undoubtedly cause the attentive reader to ask: What security can be expected from a vault door that opens automatically in the morning in the absence of the bank manager?

And quite rightly so! Should the bank manager's arrival be delayed at the critical moment then the safe would be accessible to anybody. This danger has been neutralized by installing behind the main vault door some lighter doors fitted with combination locks.

Before concluding, the writer wishes to acknowledge his indebtedness to the late Mr. John M. Mossman, prominent banking architect of New York and to the late Mr. Charles Courtney, who was for so long President of the Master Locksmiths Association, of New York, for their valuable information and suggestions, based on technical developments in the U.S.A.

While on an instructional tour through the U.S.A. in 1902 the writer had the privilege of making the acquaintance of Mr. Mossman and the many days spent with him in his Museum, which contained old safe and vault door locks collected by Mr. Mossman during his career as a banking expert, are still fresh in his memory, as are the recollections of his unceasing efforts to develop the writer's knowledge of security locks for safes and vault doors of every description.

Mr. Mossman was a recognized bank architect and adviser and the writer was offered every opportunity of profiting from his wide practical experience in the field of bank security devices including vaults, safes and safe deposits. Thanks to his introduction I was admitted to many lock factories and to quite a number of banks in places covered by my tour. To my pleasant surprise I received, about two months after my return to Holland, a number of bulky and heavy cases, containing a choice collection of bank locks, duplicates of those in Mr. Mossman's collection.

These formed the foundation of the present Lips' Collection, and whilst this occurred over 55 years ago, a regular exchange of European and American locks and experience has taken place ever since, not forgetting the lively and vivid correspondence maintained for many years. In addition, I had the pleasure of being host to Mr. Mossman on several visits to Holland, when he visited the Lips' Safe and Lock factories. On the 6th March 1912 this unforgettable friend, to whom I owe so much, died. He bequeathed his unique collection of old locks and keys to the General Society of Mechanics and Tradesmen of New York, and created a fund of $20,000 for its maintenance.

Mr. Charles Courtney of New York was a contemporary lock expert and was Chairman of the International Association of Master Locksmiths, of which I became an honor member on the 10th June 1935. The ties of friendship with Mr. Courtney were laid many years ago and ever since a regular exchange of ideas and experience has been maintained with his successors. Mr. Courtney was also a welcome visitor to the Lips' Works in Dordrecht on several occasions.

282. Smaller type vault door with Automatic Device in the Unlocked Position.

283. Smaller type vault door with Automatic Device in the Locked Position.

Much to my gratification a mutual exchange of American and European inventions and constructions still continues.

I sincerely hope that I have succeeded in arousing the reader's interest in the great variety of security devices applied throughout the ages, passing through the different stages, from the primitive way of safeguarding property by hiding it in a hole in the ground, covered by a stone, up to the present systems of protecting property by means of highly ingenious and refined security devices, comprising strong boxes, safes, safe deposits and bank vaults.

The evolutions have led to the creation of a specialized industry in many countries, employing many thousands of workers.

I am proud to have contributed something to the development of this particular industry in the Netherlands, Belgium and Italy and look back on my 58 years' activity in this field with satisfaction.

This book is concluded with an illustration of a Lips' circular vault door, a culmination of present day technique, being one of the strongest and heaviest Strong Room Doors in Europe.

Antique Locks and Keys:
Their History, Uses and Mechanisms

Author: Ulf Weissenberger

ISBN: 978-0-9979798-9-3

Spring 2020

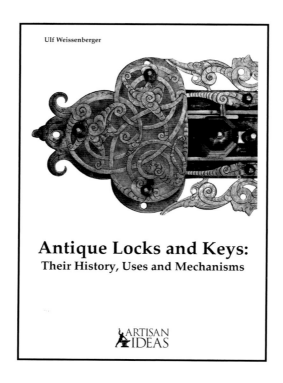

For over 25 years the author, Ulf Weissenberger, has been collecting and restoring antique locks and keys. This extraordinary book is based on his experience and features many of the most interesting pieces from his extensive and impressive collection.
In this large sized hardcover Ulf Weissenberger's text and explanations are accompanied by over 850 high-quality, full-color photos and illustrations.

From the Roman period and even earlier, up through the nineteenth century, the author explores, with great attention to detail, a wide variety of locks and their keys which were used for various types of chests, doors, cabinets, jewelry boxes, and more.

An important feature of this book which locksmiths, collectors and other readers will greatly appreciate is that the locks' internal mechanisms are shown in crisp, high resolution photos – as are the keys.

Hardcover, 12-1/3-in x 9-1/3-in x 1-in, 850+ photos, 288 pages.

Order from **www.ArtisanIdeas.com**, or your preferred retailer.

Locks from Iran:
Pre-Islamic to Twentieth Century
(Updated and Expanded Edition)

Author: Parviz Tanavoli

ISBN: 978-1-7333250-1-1

January 2020

In Iran, the padlock was developed in an amazing variety of sizes, shapes, materials, and mechanisms, the likes of which are less frequently encountered in Europe. On the whole, very little attention has been devoted to the history and development of the lock in Iran. Parviz Tanavoli, one of Iran's leading sculptors – and lock collectors – was first attracted to the locks of his own country because of their sculptural qualities.

In this beautiful and fascinating book the author shares with us the most interesting examples of locks from his famous collection which was first introduced to the American public by the Smithsonian on the occasion of the American Bicentennial.

This first edition of this well-illustrated book was published by The Smithsonian Institution in 1976. In 2019 the author decided to update the original book adding new text and a great number of new high quality photos. In addition, all black and white photos which appeared in the original edition have been substituted with full color photos.

Hardcover, approx. 160 pages, hundreds of full-color photos and illustrations.

Order from **www.ArtisanIdeas.com**, or your preferred retailer.

The Spruce Forge Manual of Locksmithing:
a Blacksmith's Guide to Simple Lock Mechanisms
(Completely Revised and Updated 2nd Edition)

Authors: Denis Frechette and Bill Morrison

ISBN: 978-1-7333250-0-4

Spring 2020

This is an updated and revised edition of one of the favorite manuals for blacksmiths, locksmiths and lock enthusiasts in the USA. It is also the only book of its kind available. Unlike most locksmithing books, which deal solely with parts replacement, this book illustrates the techniques and tools utilized in the making of hand forged locks.

The original projects in the previous edition have been revised and new projects have been added.

Illustrations, high-quality color photos, original lock patterns, and step-by-step instructions guide the reader through the process of traditional lock making.

Hardcover, approx. 130 pages, hundreds of illustrations.

Order from **www.ArtisanIdeas.com**, or your preferred retailer.